First Technology

R.V. SCRINE and E. CLEWES

SERIES EDITOR – DAVID SHAW

HODDER AND STOUGHTON

LONDON SYDNEY AUCKLAND TORONTO

ACKNOWLEDGMENTS

We would like to express out thanks to David Shaw whose efforts have contributed a great deal to the promotion of craft, design and technology. His guidance proved invaluable during the production of this book.

The authors and publishers wish to thank the following firms for permission to reproduce copyright photographs in the pages of this book:
NASA, p7; Zanussi, p15; Cavendish Cash Registers, p20; Casio Electronics Ltd, p20; Tiffany & Co., p 26; A Shell Photograph, p36; Caradon Terrain, p37; General Motors, p45; North of Scotland Hydro Electric Board, p48; Gaucho Leisure Ltd, p50; Black & Decker, p54; Gyproc Insulation Ltd, p55; Strathclyde Passenger Transport Executive, p 56; TI Raleigh Ltd, p58 (bottom right); Allsport, p58 (top left); Abbeydale Industrial Hamlet, pp62, 64; David Mellor p70; The National Motor Museum, Beaulieu, p74; Bernina, p75.

Series Editor: David Shaw

ISBN 0 340 41159 7

First printed 1989

Phototypeset by Tradespools Ltd., Frome, Somerset
Printed in Great Britain for Hodder and Stoughton Educational, a division of Hodder and Stoughton Ltd, Mill Road, Dunton Green, Sevenoaks, Kent TN13 2YD, by Thomson Litho Ltd, East Kilbride, Scotland.

Contents

PREFACE

Between the ages of eleven and fourteen, boys and girls should have the opportunity to try a wide variety of practical challenges at school. This book aims to provide challenges of a technological nature. 'Science' and 'craft' can no longer be separated and this book provides a bridge between them.

At the end of the book is the Service section which is intended to provide back-up information to assist mainly with practical work. The index has been provided with the intention that it will be of particular use to pupils researching projects.

The illustrations and words have been made as clear as possible so that pupils can make progress on their own. Teachers should find that this will help to provide a basis for individual work, homework and work in areas without specialist equipment.

We think that First Technology will also give pupils a good introduction to craft, design and technology subjects at examination level. Pupils who do not take CDT examination courses will have tried a variety of problems which will help them to see science from a practical point of view.

Introduction

How did you get to school today? Did you walk, ride a bike, get a lift, come on the bus or train? At one time the only way to get from one place to another was by walking. Look at how many choices that we have now. What has happened to make it possible to have such a choice? The answer is technology.

Technology is not new; as soon as the first man used a piece of wood to move a heavy object he had become a technologist.

A wheelbarrow would make it easier to carry heavy loads. A horse pulling a cart would reduce the work for the farmer.
A steam engine could pull greater loads than the horse and a tractor is smaller and more convenient than a steam engine.

Describe how technology has changed the food that people eat. What advantages and disadvantages have resulted from technology? Use the following in your description:

- bow and arrow
- fire
- tools
- metal
- machines
- chemicals

The distance between London and Birmingham is about 170 kilometres. Technology cannot alter the distance between cities but it has reduced the time needed to travel from one to another. Using the information overleaf work out how long it would take to travel a distance of 170 km for each method of travel. You will probably need a calculator.

You would have to stop and rest on some journeys which would make them take even longer.

Methods of travel have improved as a result of using scientific ideas in practical ways. Our lives have become dependent on the ways in which science can be used. Technologists are people who can put science to work for us.

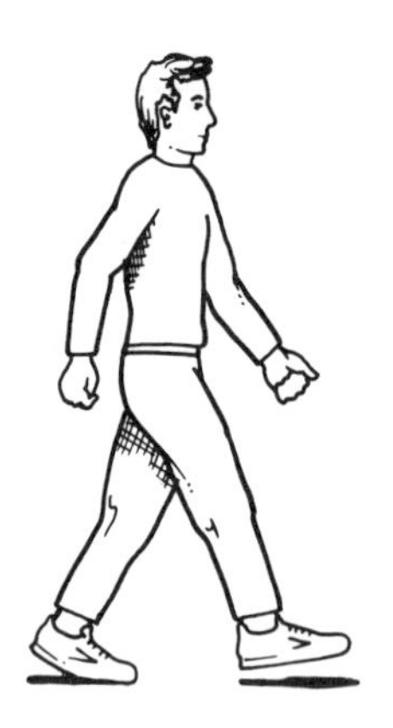

walking speed of 5 kilometres per hour (km/h)

a road racing cyclist at 40 km/h

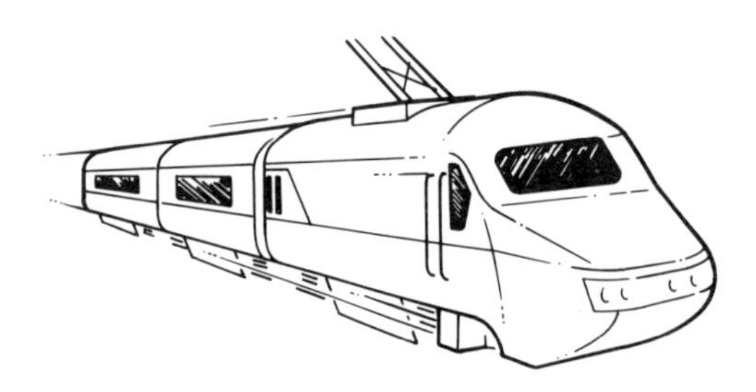

an electric train at 145 km/h

a fast mail coach, pulled by horses in 1830, 22 km/h

a motor car at 112 km/h

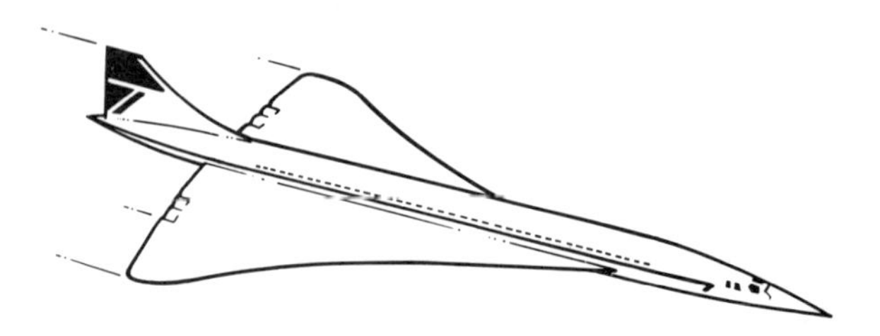

Concorde at mach 1 (1226 km/h)

The information about travelling times could be shown on a bar chart like the one below.

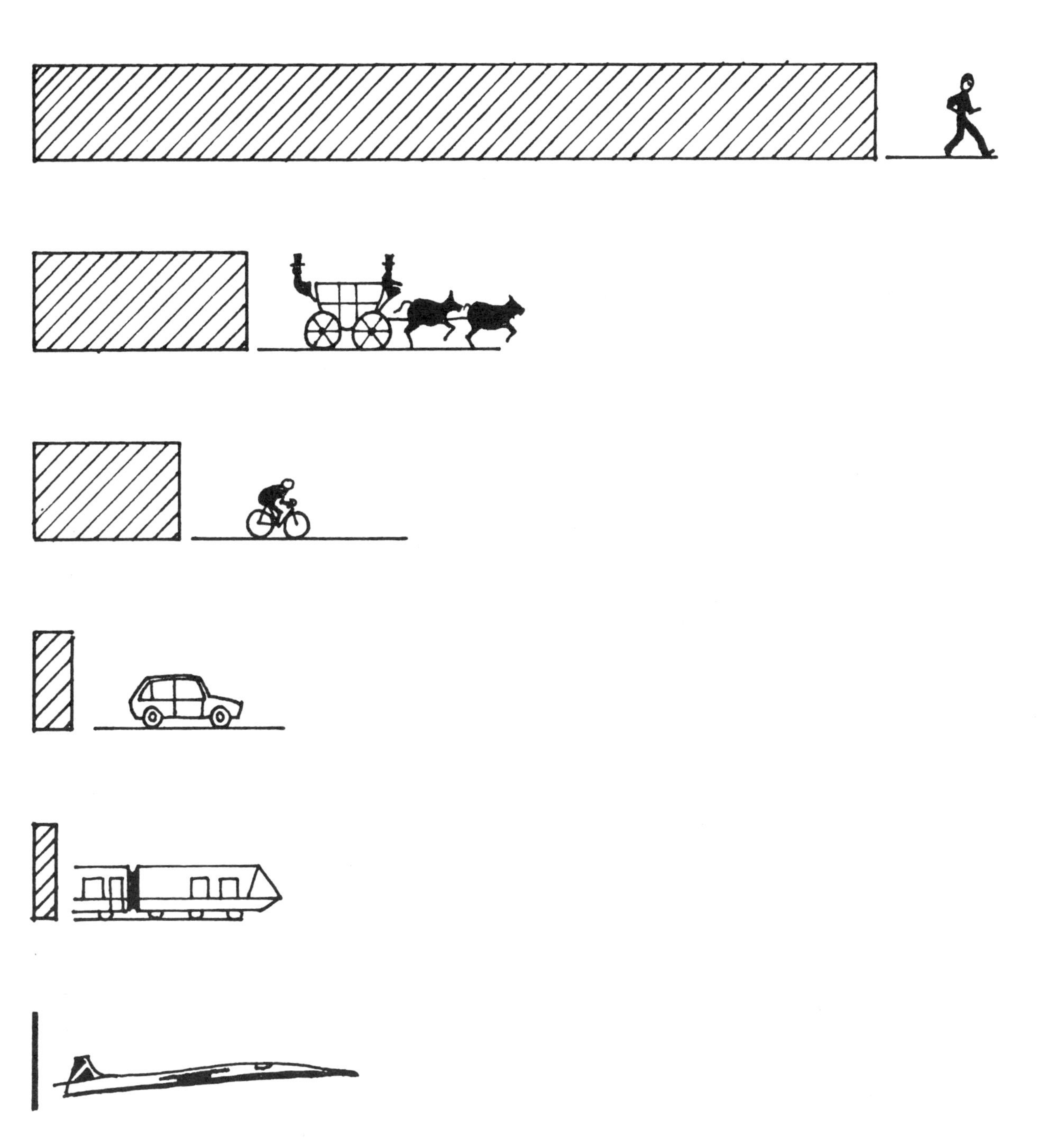

Time taken to travel 170 km

The bar chart on page 7 represents the answers to your calculations

If you could keep the speed constant (not changing) it would take 34 hours to walk, 7 hours and 42 minutes by stage coach, 4 hours and 12 minutes for the racing cyclist, 1 hour and 30 minutes for the car, an electric train would take 1 hour and 12 minutes and Concorde would take 7 minutes.

Use the information given about transport and display it in another way which shows the differences clearly.

You could use colours or make a three-dimensional drawing (one that looks solid).

Using 'sending a message' as a theme (idea), find information that could be used in the same way as the travelling times shown.

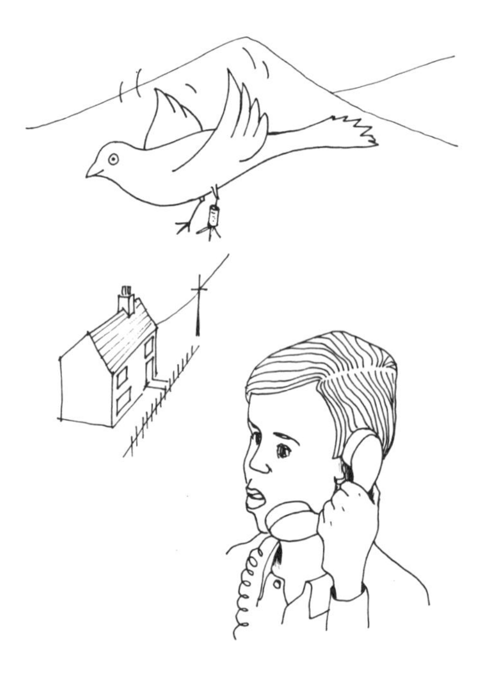

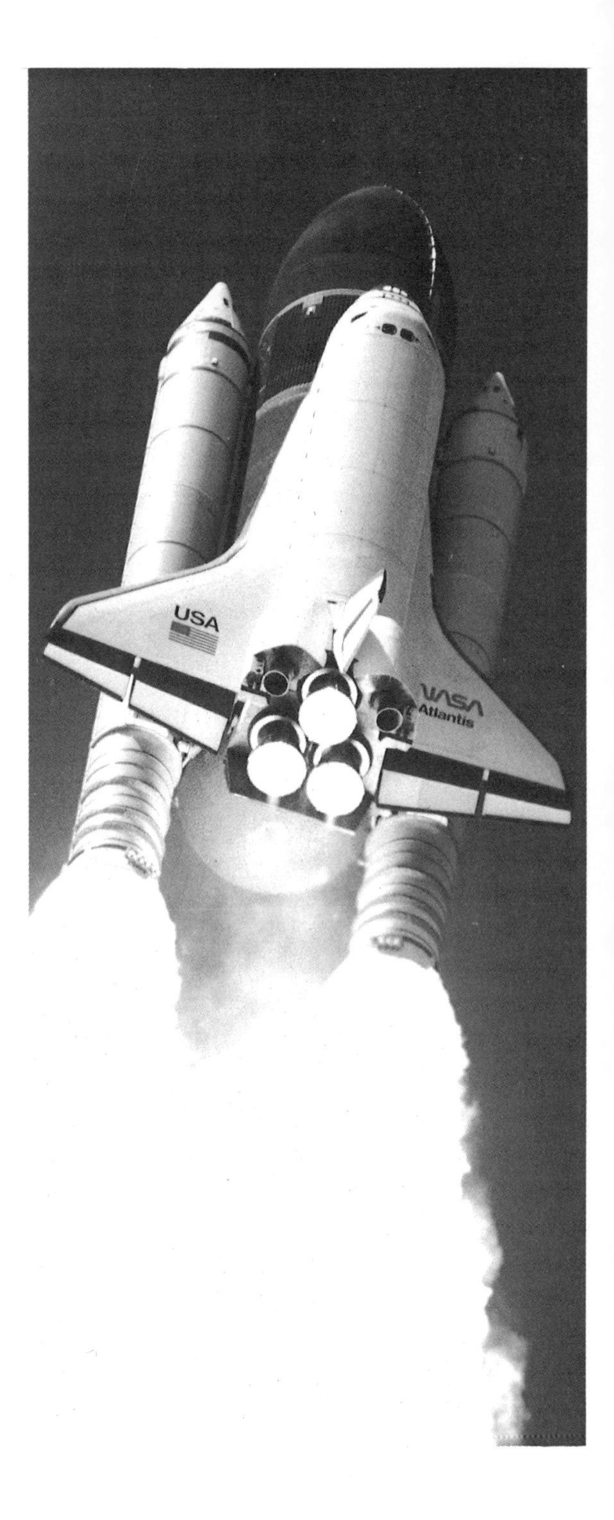

People's travel into space began in 1961. None of the early space vehicles were re-usable. Electricity made sending messages faster. Similarly, the first space-shuttle in 1981 speeded up the space programme by saving months of time taken to build new vehicles.

The 'Voyager' spacecraft carry information about us; pictures and music out into the universe. Space travel can also be a way of sending messages even if we do not know who will receive them.

Space travel is the most advanced form of travel that we know today. How will your children get to school?

Make two lists showing the advantages and problems of space travel.

Exploring and explaining ideas

DESIGNING AND MAKING

There are different kinds of design. An artist may design a new wallpaper pattern.

A politician may have designs for where a new motorway should go.

An engineer may design a motor car to run on the motorway.

All these designs mean 'having ideas which can be made in some way'. There are a number of steps which must be taken between the first idea, and the finished product. This is called the design process.

THE DESIGN PROCESS

Step 1: Situation

A problem may affect many people or just one person.

Step 2: The brief

The brief is a statement of what is to be designed. Write down this statement clearly.

What prompted the designer to design a shopping trolley in the first place?

Step 3: Ideas and research

Discuss an idea or ideas within a small group. This is called 'brain-storming'.

Discussing a problem in this way can start an idea that one person alone, might not have thought of. It might also help you to see a problem in a different way.

Share your ideas with your teachers and your parents, as well as with other pupils.

Can you think of other ways of getting ideas, apart from brain-storming?

Step 4: Ideas sheet

Make sure that you record all your ideas on:

(i) paper, by way of sketches and notes
(ii) on tape (spoken word)
(iii) photographs

Sketches with notes alongside them are referred to as 'annotated sketches'. Sketches need not be the best quality at this stage. A tape recorder can be very useful to record your first thoughts.

Taking photographs of your drawings, models and sketches can form a permanent record of your work.

What would be the best method of recording ideas for the shopping trolley?

Step 5: Chosen idea

By the time you have completed your ideas sheet, you will probably have decided which idea you will develop.

An ideas sheet usually develops in the following way:

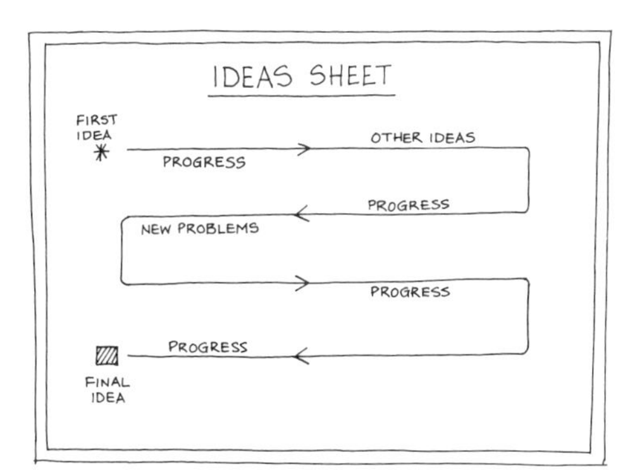

The diagram shows idea(s) developing as you go along. Sometimes, it can happen, that you choose an idea between the * and the ■. Make this chosen idea clearer by making a pictorial sketch. You will find more about how to do this on page 14.
Why do you think it is more likely that your final idea is the one you will choose to develop?

Step 6: The model

A useful way by which you can show what your chosen idea will look like, is by making a scale model. A scale model is one which is not full-size. A model lets you see an object in three dimension (3D), so you can easily check the form (shape), height and width etc—A model does not have to be made of the same materials as the finished article.

You could make scale models of surrounding objects and persons, to get a better 'feel' of the design.

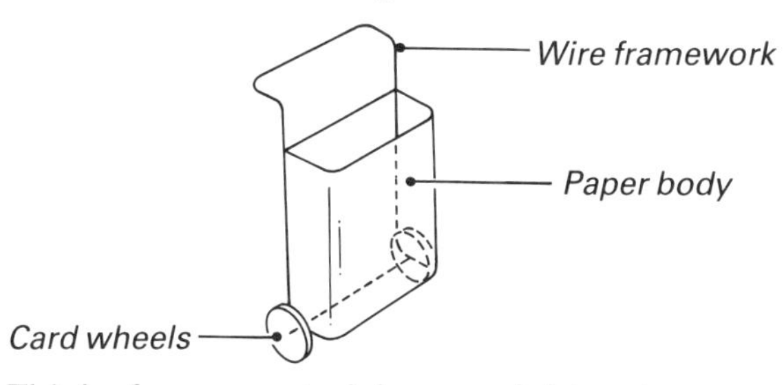

Think of some materials you might make your model from.

Step 7: Working drawing

A working drawing is necessary so that you can make your article to the correct sizes. It will show three or more views e.g. a front view, a plan view and an end view, as shown in this diagram.

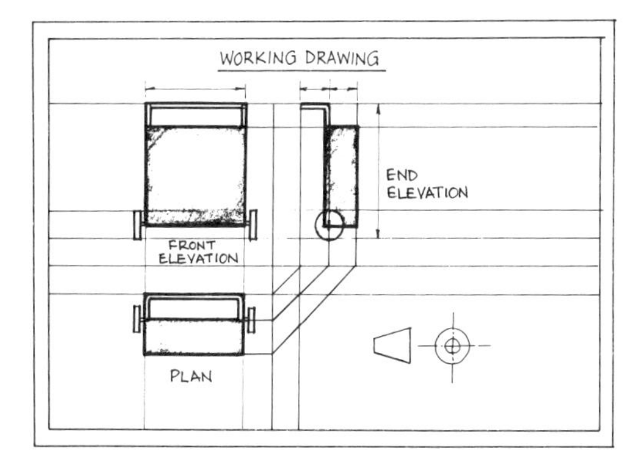

Your design is drawn less than full-size (scaled-down) but with the actual measurements in position.

A working drawing has to be drawn very accurately. Can you think why?

Step 8: The prototype and test

When you are satisfied that the model looks right, you should make a prototype. A prototype is a full-size working version of your chosen idea. It can be properly tested to see if it will really do the job for which it was intended.

Any changes or alterations are made to the article after testing the prototype. These changes are called modifications.

Diagram to show the design process.

The arrows show the direction of your thinking. They will often work both ways, because you will find yourself referring back to the previous stage to check whether or not or not you are working along the 'right lines'.

What name do we give to the type of diagram shown below?

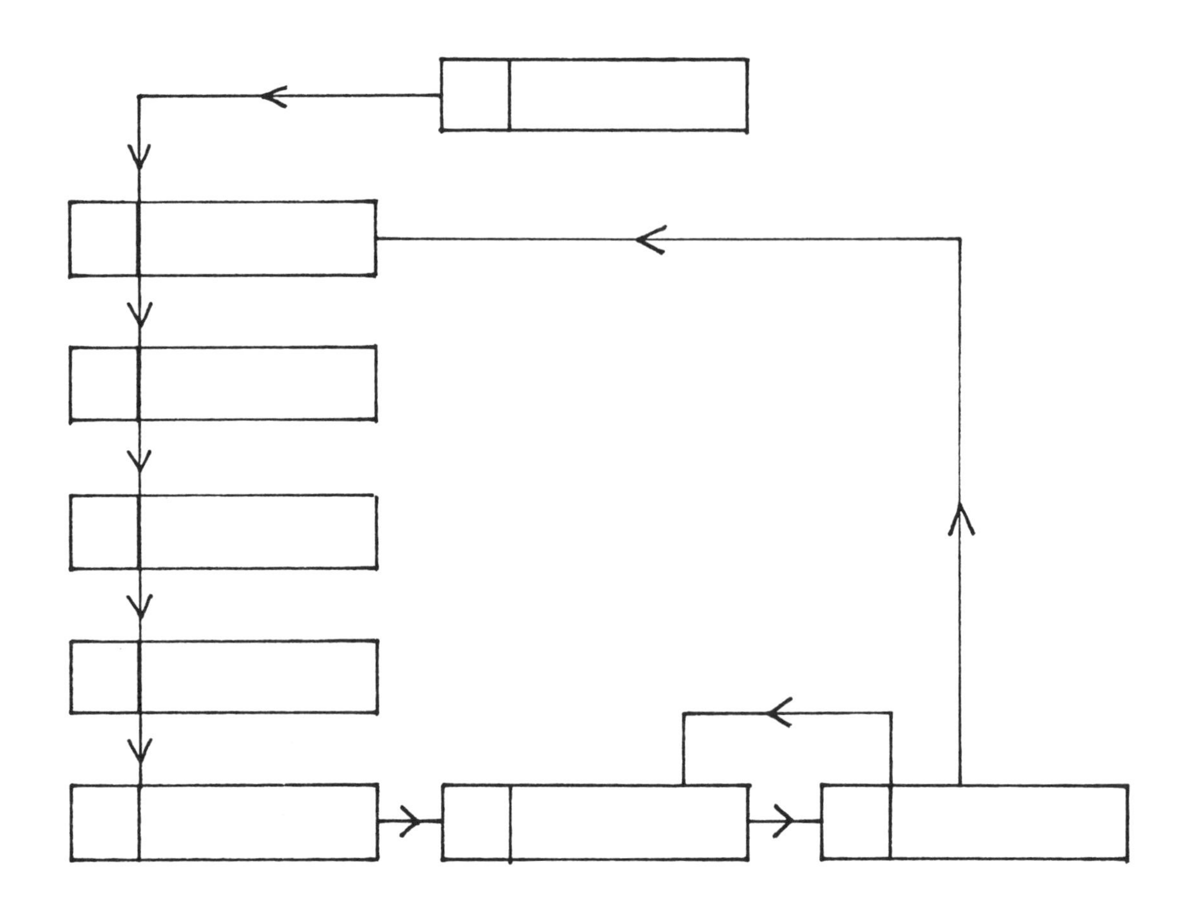

Conclusion

It is always a good idea to look at other people's designs, particularly of those who earn their living by making 'ideas work' such as industrial designers, architects and interior designers.

Always be on the look-out for ideas which may start a new train of thought. Discuss your ideas. Make notes and visit galleries and exhibitions, where you will see other people's designs.

When faced with a design problem, ask yourself:

'What is the finished article expected to do?'

'How will it be used?'

When evaluating your completed design (checking to see if it is a success), ask yourself:

'Does it work well?'

'Could it work better?'

'How could it be improved (modified) to make it work better?'

'Could it be made more cheaply?'

Finally, here are some situations which need solutions.

Try them for yourself.

USING THE DESIGN PROCESS

1 A number of cotton reels are heaped in a box in an untidy manner. They need to be placed in a position where they are easily and quickly identified.

2 A person in a wheelchair has difficulty in using ordinary garden tools and has asked you to design better ones. What questions would you need to ask before starting? Write a list of things that you think would be important.

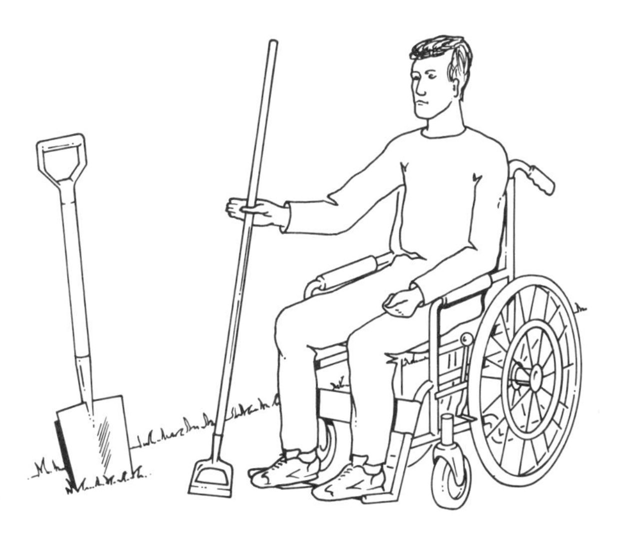

3 A bike has to be leant against the wall because there is no other means of supporting it. Devise a method of supporting it so that it can stand away from the wall?

4 Cassettes are placed on a shelf in a jumble and need to be stacked in an orderly way so that they can be easily read, identified and picked up.

When thinking about this problem you could look at other people's answers that are for sale in shops. They will have advantages and disadvantages. List these and use them to improve your design.
Why is it that often you see things for sale in shops that are poorly designed?

Making freehand drawings

TYPES OF PENCIL AND THE LINES THEY WILL PRODUCE

On a new pencil, at the end opposite the point, are two letters or a letter and a number. This code helps the designer, draughtsman or artist to make the correct choice of pencil, for a given piece of work. The chart below will help you to choose the best pencil.
Copy out this chart for yourself and experiment with patterns of lines, using (if possible) a variety of pencil types.

PENCIL TYPE H GRADE								(H	B) PENCIL TYPE B GRADE								
9	8	7	6	5	4	3	2	1	1	2	3	4	5	6	7	8	9
← The higher the number – the harder the lead. This type of pencil is used in the technical drawing room to make very precise drawings. Be careful not to press too hard. It will make a ridge in the paper.									→ The higher the number – the softer the lead. This type of pencil is used in the art room for sketching and shading. Because the lead is soft, it can fall onto your paper in the form of dust. That is why your drawings are sometimes grubby-looking after an art lesson.								
EXPERIMENTING WITH H AND B PENCILS																	

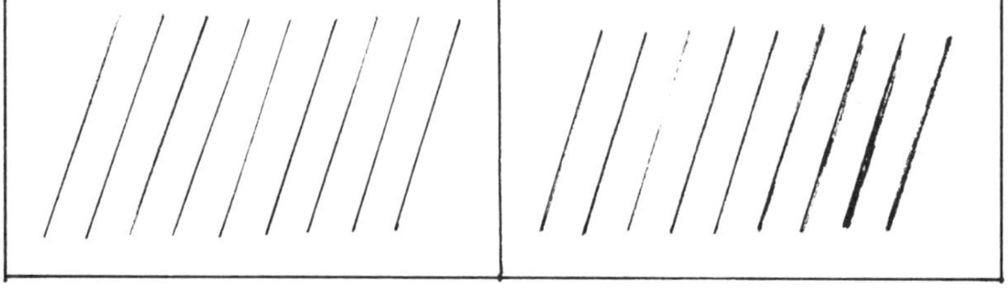

Look at this page. It shows how these pencils can be used to draw an everyday article.

Types of line

This type of line is used when constructing a drawing.

(a)

This type of line is used to present a definite finish or outline to the drawing.

(b)

Examples of when to use these lines would be:

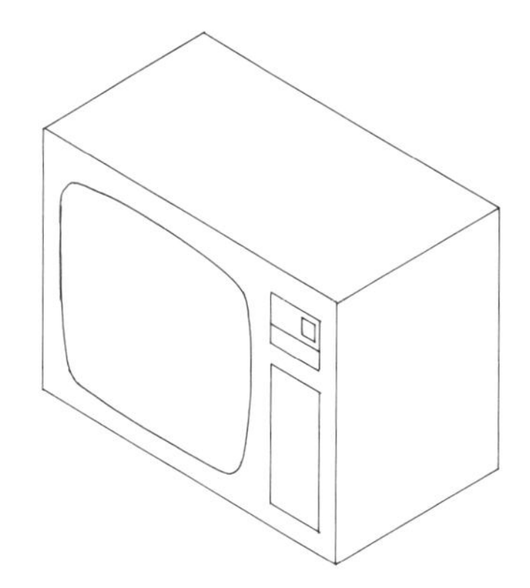

Check that it is correct and then complete the drawing.

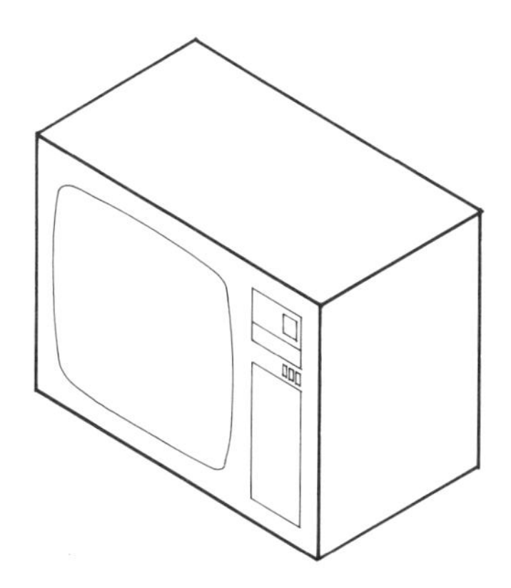

Try not to rub out your guide lines. They should fade into the background when the drawing is completed.

Occasionally, a much darker, heavier line can be used to make the outline stand out. This type of line is produced, using a broad felt tip pen. Be careful not to make the outline too heavy. It can be overdone!

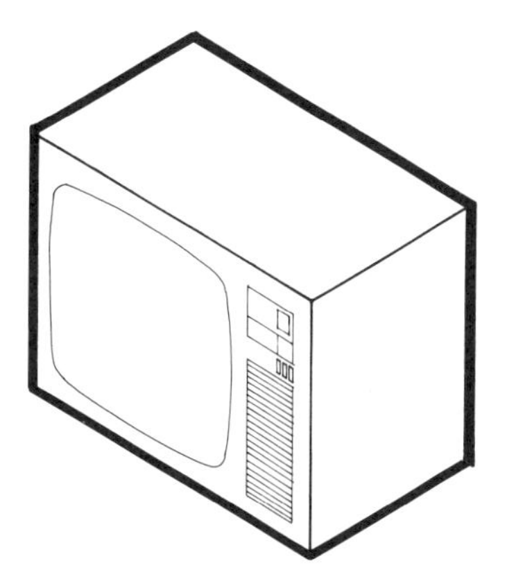

Two objects that could be drawn using this method

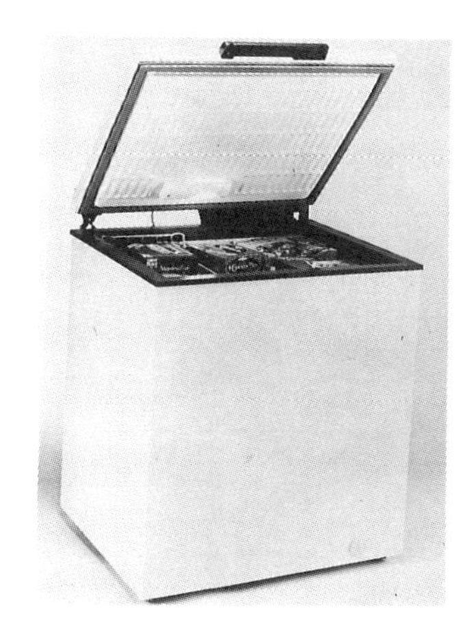

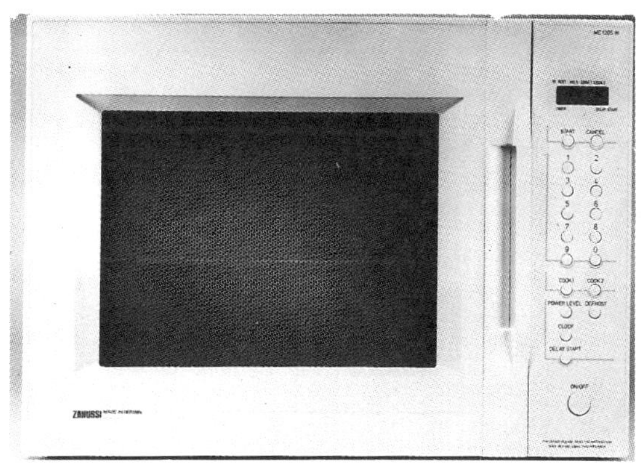

AIDS TO DRAWING

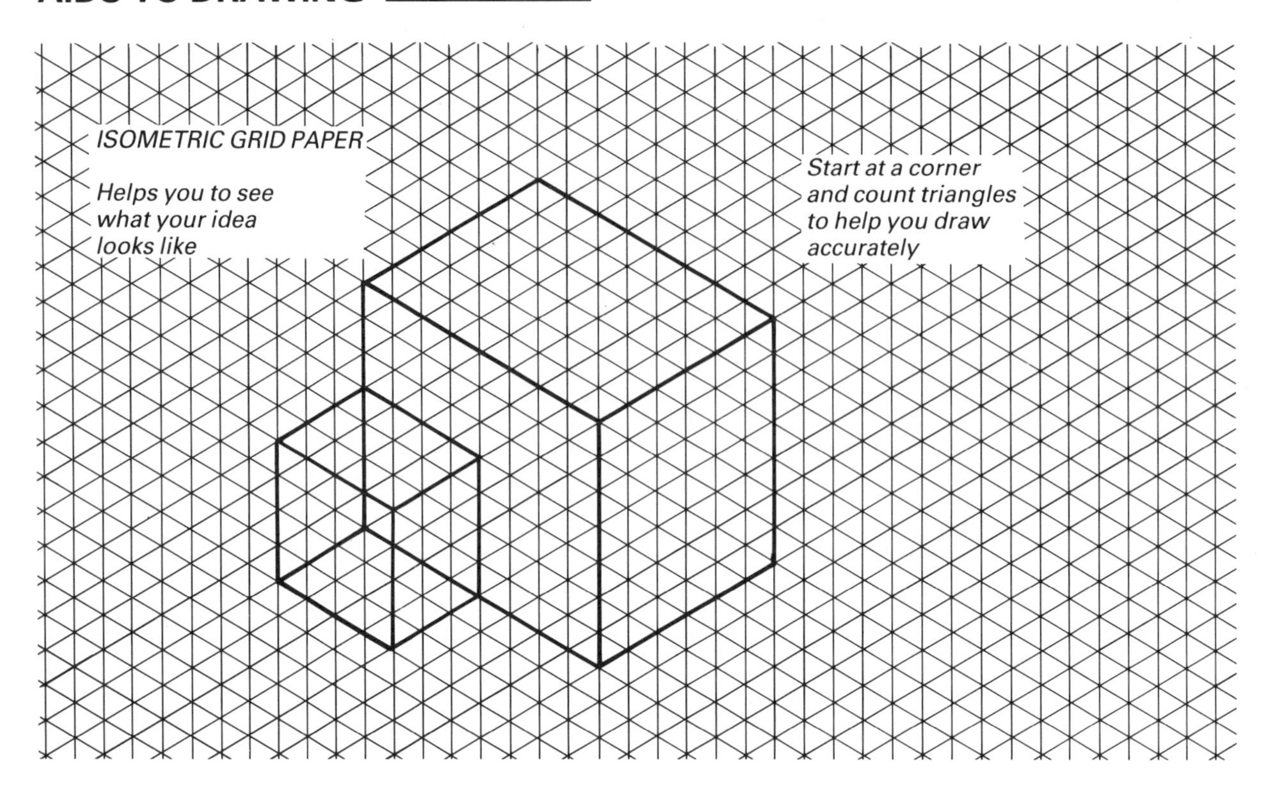

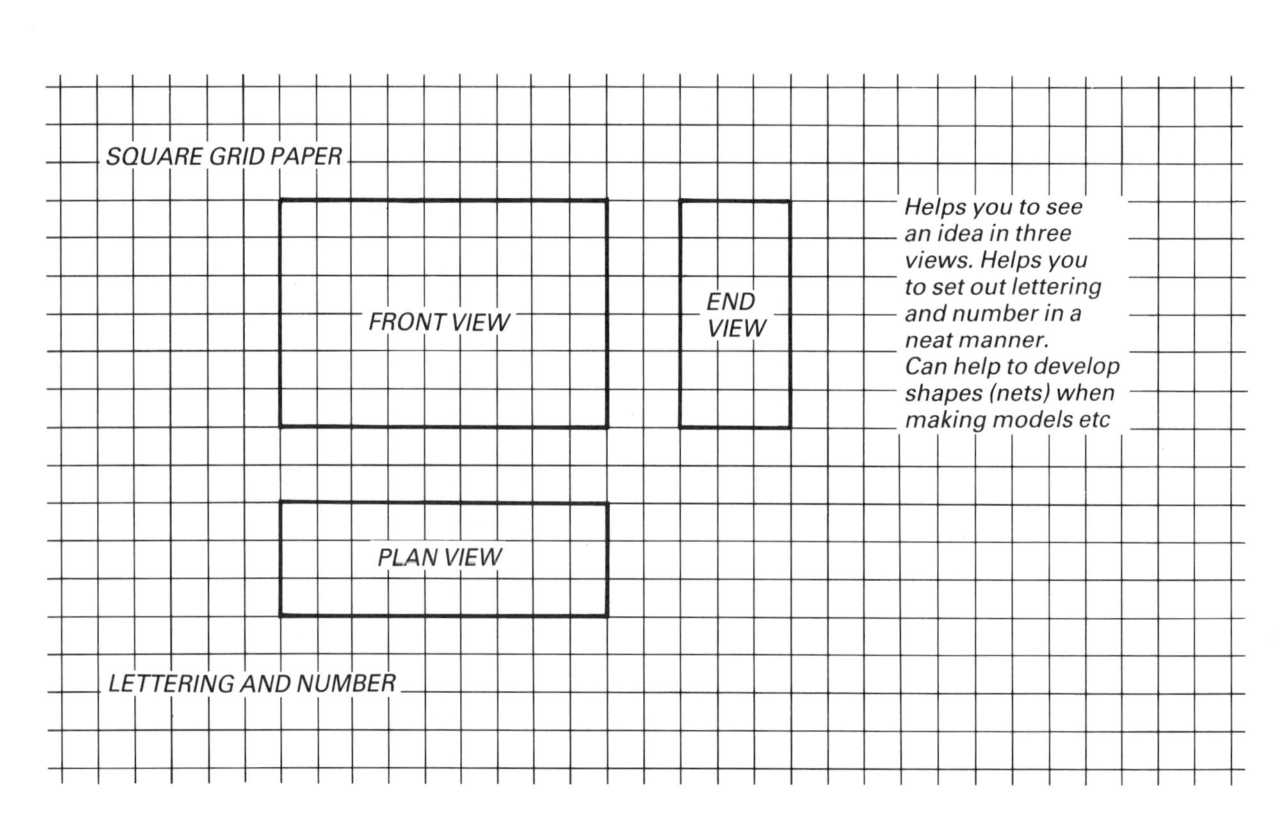

USE OF LIGHT AND SHADE

Good, accurate shading improves the look of a drawing and gives it that 3D feel. It should be carried out only on your final presentation work.
Single strokes of the soft pencil or pen will produce this effect. This method is slow and you will need lots of patience.

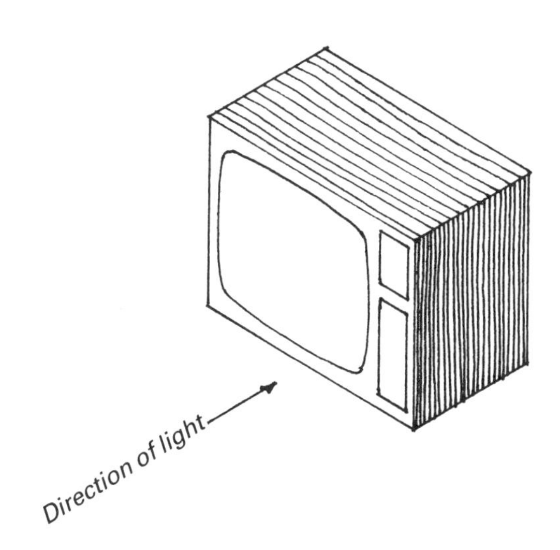

Method (i)

Rub the lead of a soft pencil (2b) into the surface area, then with the tip of your finger continue to rub it in still further.

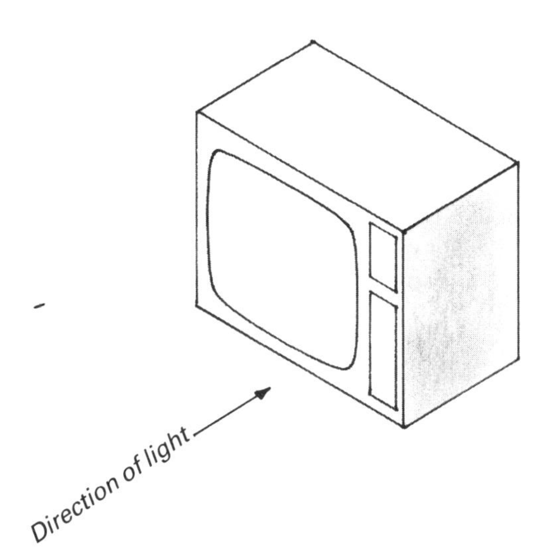

Method (ii)

Feed a piece of cotton wool with pastel crayon and rub this into the surface area, concentrate on the darkest face first.

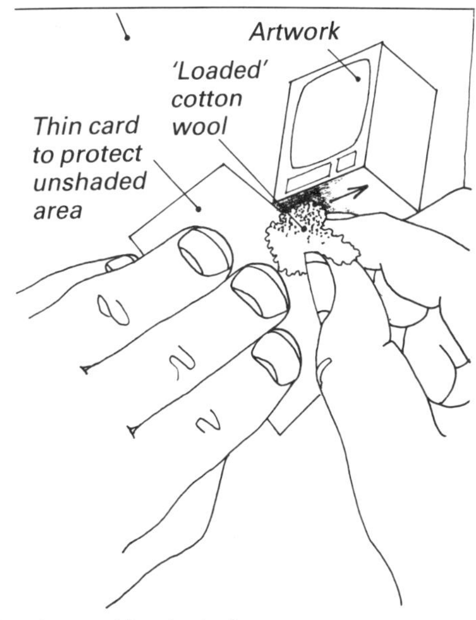

A simple masking technique

With both methods it will be necessary to mask off (cover up) the surface areas which are not being shaded. The rubbing method will produce a softer effect.

The photograph above shows three areas of light and shade

Two sets of examples using the shading methods on page 17 are shown below:

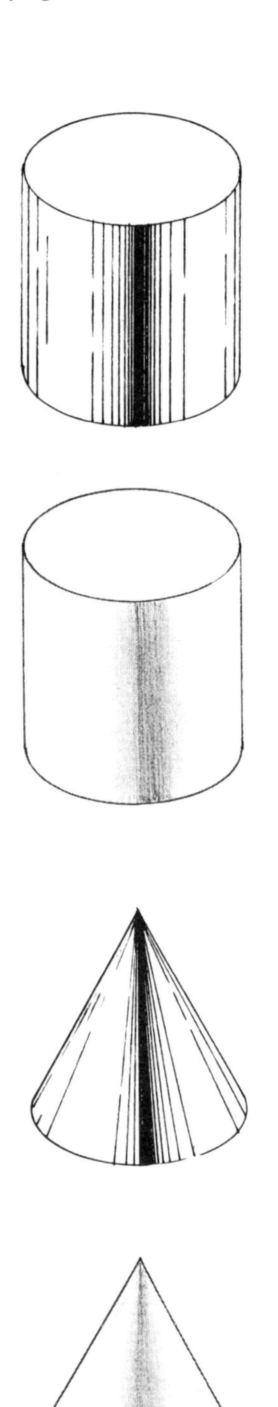

Below are three everyday articles shaded to the two methods described.
Choose some of your own and try shading them.

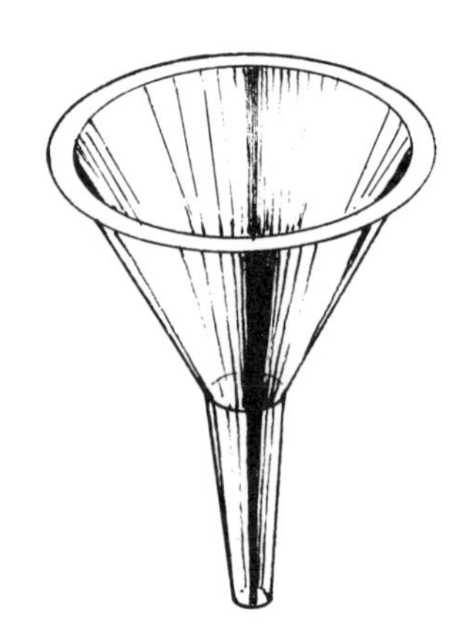

Funnel

Mug

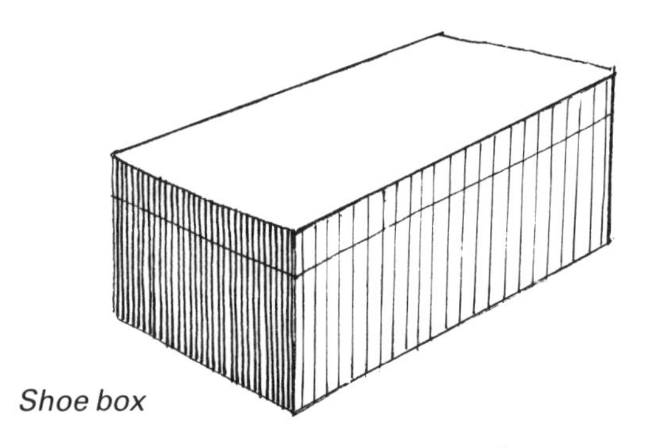

Shoe box

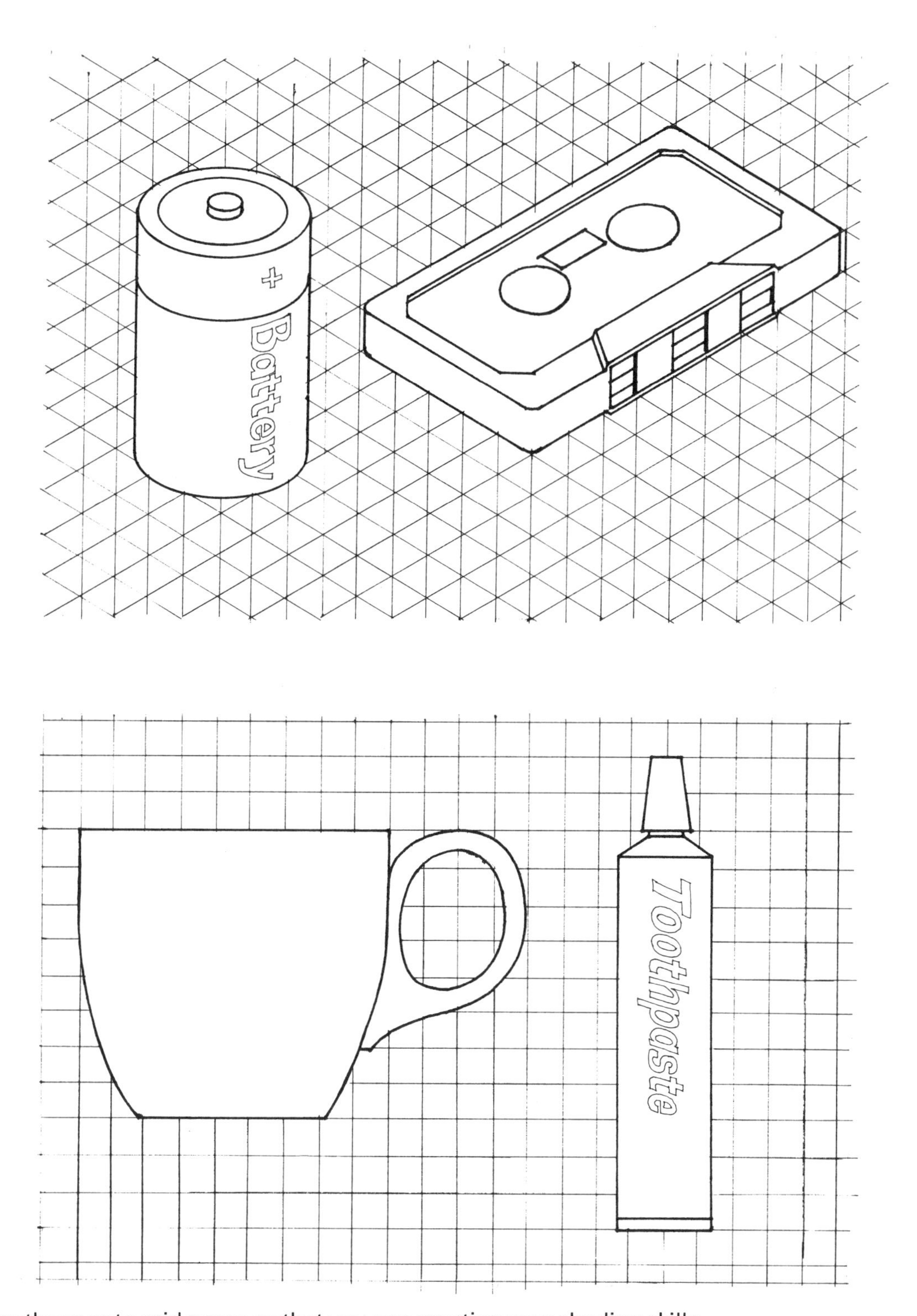

Copy these onto grid paper so that you can practise your shading skills.

The calculator—an everyday item

All of you know what a calculator looks like. There are many calculators of different designs which you can buy.
Some are made to be used on desk tops. They are usually large and clumsy.

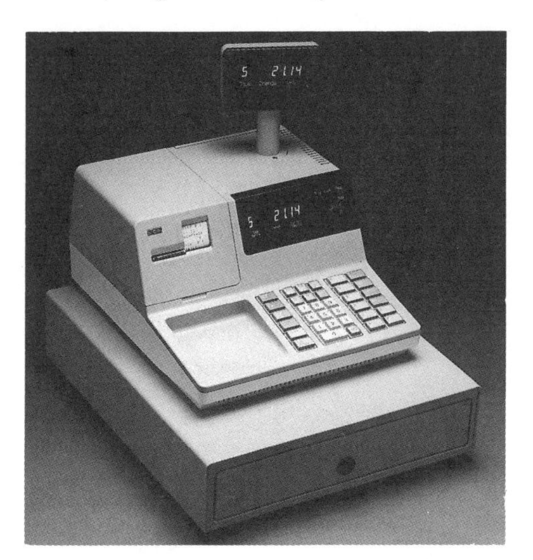

You can see this calculator being used every day at the supermarket checkout. It will provide you with a read-out on the display panel and printed read-out so that you can check the sum total against the items you have bought.
This type of calculator has large keys. Why?
You can buy a wristwatch with a calculator built-in. But some people might have difficulty using this type. Why?

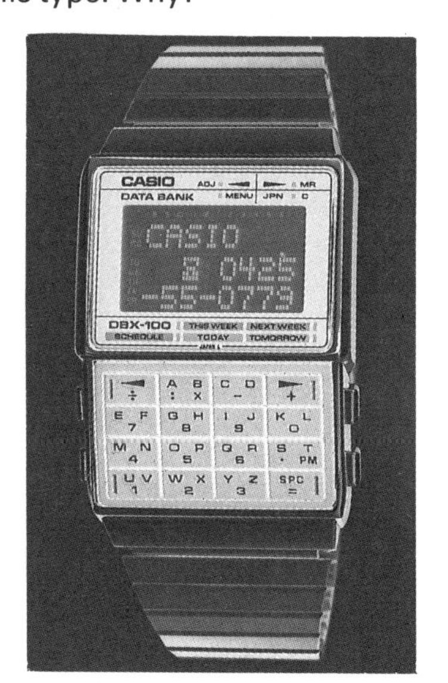

The calculator shown in the diagram is called a 'pocket calculator'. Do you know why? It is powered by either two small 1.5 V batteries or an adaptor, which can be plugged into the mains socket in your home. It has a large display window. Why? The arrangement or layout of the keys vary from calculator to calculator. Whatever the layout, the keys must be seen, read and used easily.
Design an arrangement for yourself, keeping these points in mind.

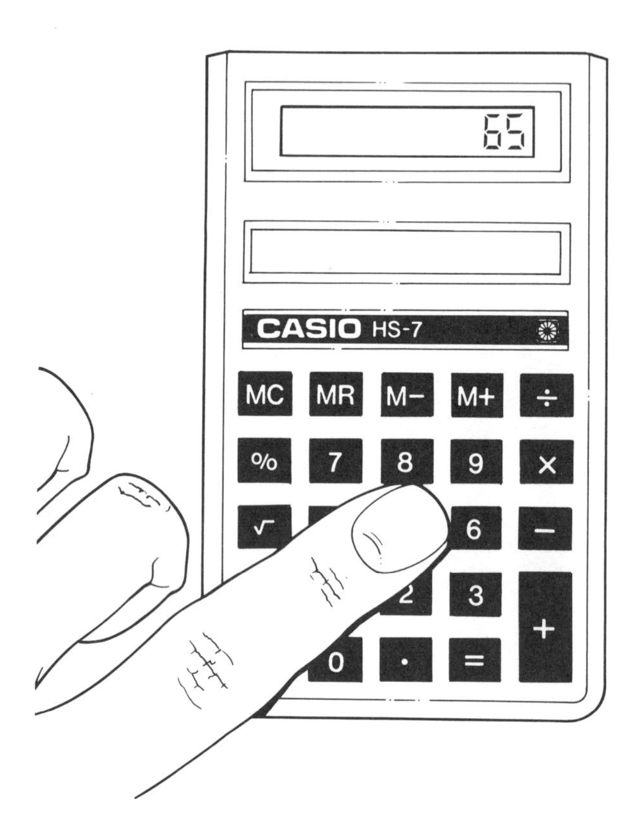

Materials

The design process helps us to solve a problem by working through a number of steps. At some of these steps you will need to decide which materials will be used. Materials have different qualities or properties and we have to decide which particular properties would be needed to solve a particular problem.

List the properties that are important in the choice of materials for the milk crate and egg mobile shown below.

An example of why the properties of a material are important can be seen in the following examples: bicycle frames are usually made from mild steel tube which is strong and not easily bent; the pipe which is most often used for carrying water around a house is made from copper which is also quite strong but can easily be bent. Copper has another big advantage over mild steel as it does not rust (or corrode) easily when it gets wet. However copper costs a lot more than mild steel.

At one time lead was used to make the pipes which carry drinking water because lead can be bent and joined easily.

Lead is not used in this way nowadays. Why not?

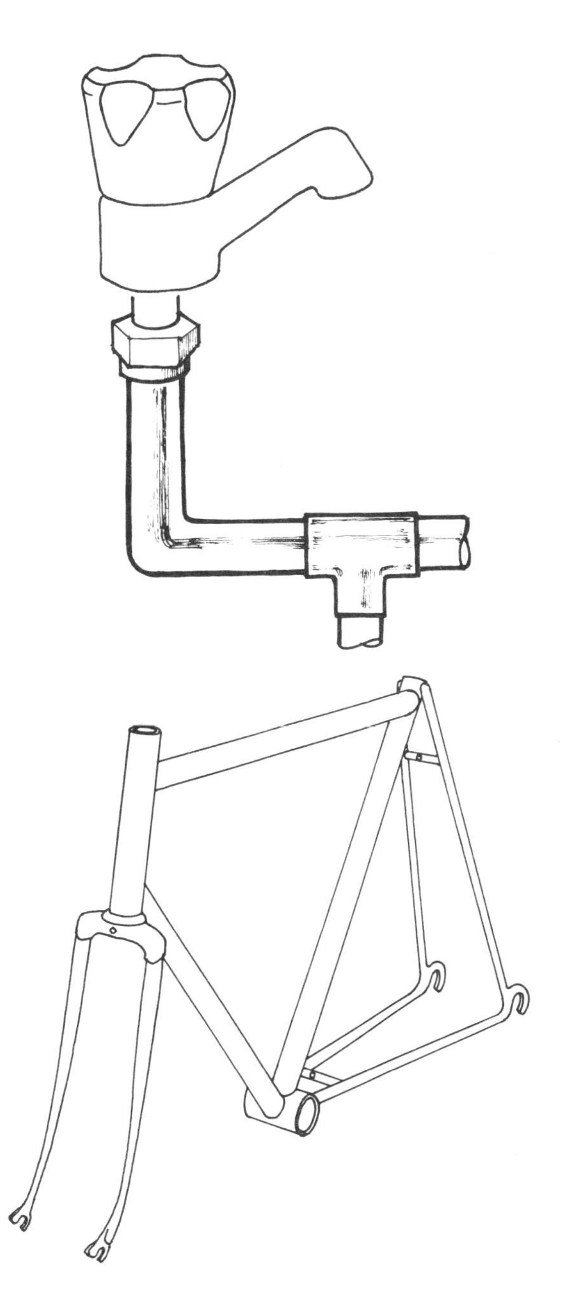

A word like 'strong' can be used when we compare the properties of different materials. For example a very hard material such as a diamond is harder than wood and wood is harder than modelling clay. The most common properties and qualities, with their opposites, are described in the following list.

Tensile strength

How easy is it to pull the material apart? Materials with a high tensile strength do not break easily when stretched. Guitar strings and fishing line have a high tensile strength. Plasticine has a low tensile strength, it is weak when stretched.

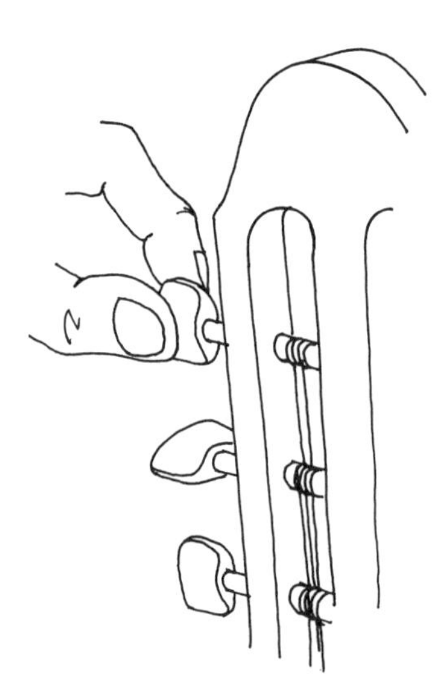

How could you compare the tensile strength of different sewing threads such as: cotton, nylon and terylene? One way would be to fasten them all to a frame and increase the weight suspended on each until the samples break.

Make sure that when the threads break there will be nothing damaged by the falling weights.

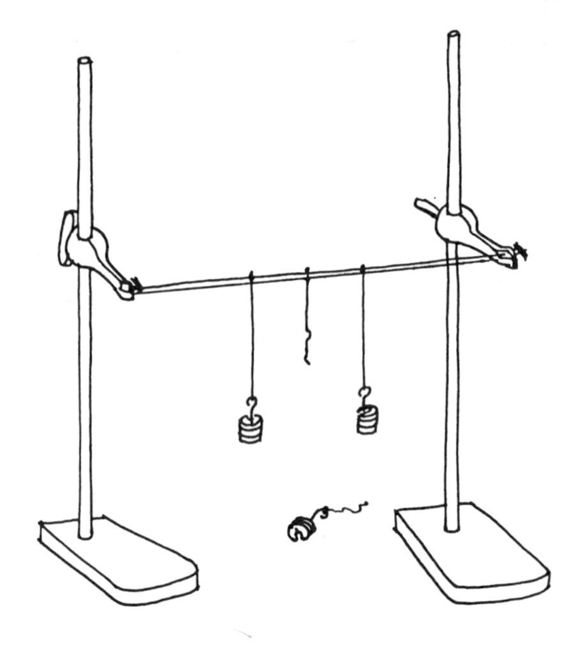

Which breaks first and which second? The first one has the lowest tensile strength.

Can you think of another way to compare the tensile strength of these threads?

Why is tensile strength important to fishermen?

Compressive strength

How well does a material resist being crushed? Concrete has a high compressive strength but expanded polystyrene does not.

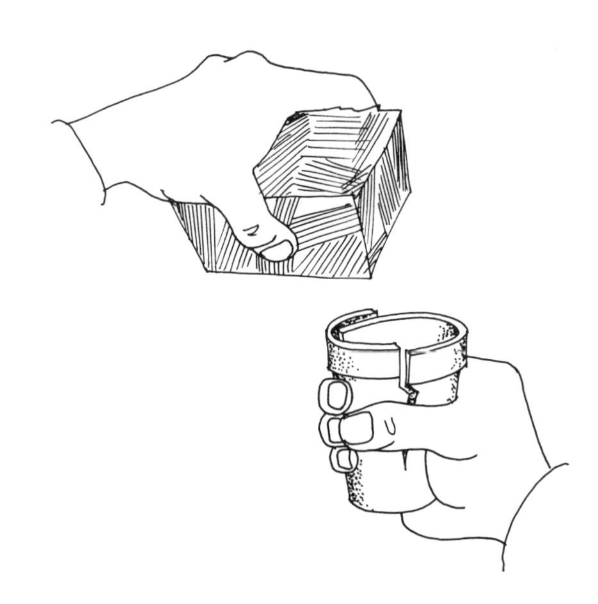

Make a pair of stilts by tying strings to empty 'cans' as shown in the diagram. The metal in the cans is very thin but the cans do not collapse easily when you stand on them. Also make cardboard stilts of the same size and thickness as the metal ones. What happens when you try walking on them?

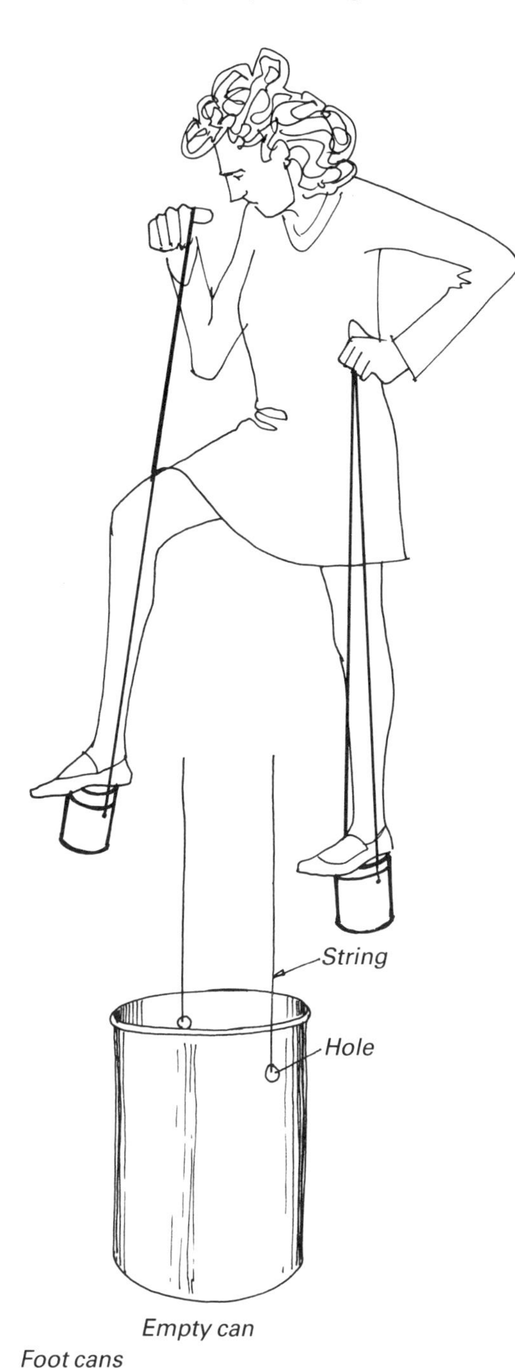

Foot cans

How could you make the cardboard foot cans have the same compressive strength as the metal ones?

Hardness

Knives and razor blades need to be very hard so that they stay sharp and cut easily. Glass and most metals are hard because they are difficult to scratch or dent. The opposite to hard is soft. Wet clay, balsa wood and cheese are soft.

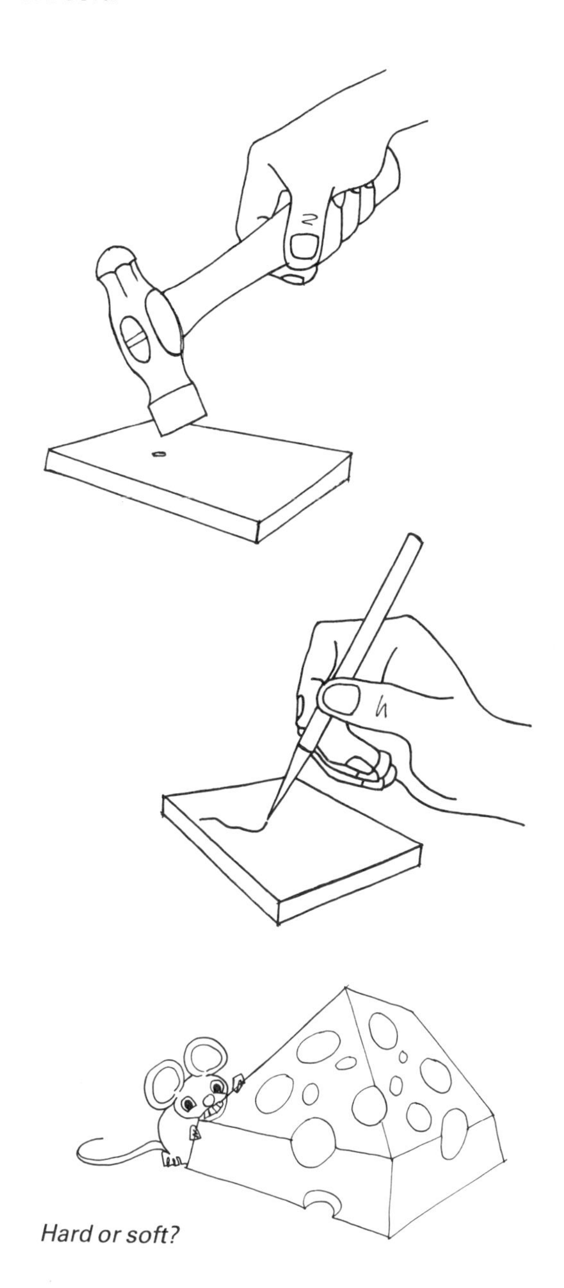

Hard or soft?

Here is one way to compare the hardness of different materials. Drop the hammer from the same height for each test. Compare copper, lead, wood and expanded polystyrene. Measure the distance across the dent made in each one. Design and make a tester using a tube with a weight which will slide inside it and a ball bearing. When using either tester do not put your fingers under the falling weights.

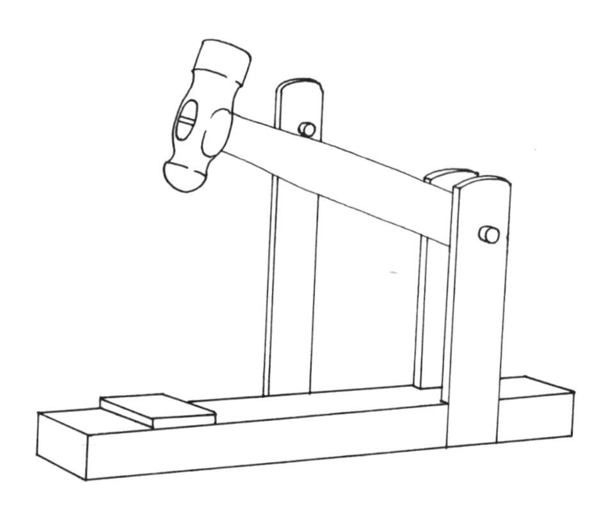

When testing brittle materials always wear eye protection and beware of flying pieces of material.

Toughness

Does a material break easily when you hit it or drop it? A polypropylene (plastic) dustbin is tough, so is soft toffee.

The opposite to toughness is brittleness. 'Seaside rock' and pencil 'lead' are brittle.

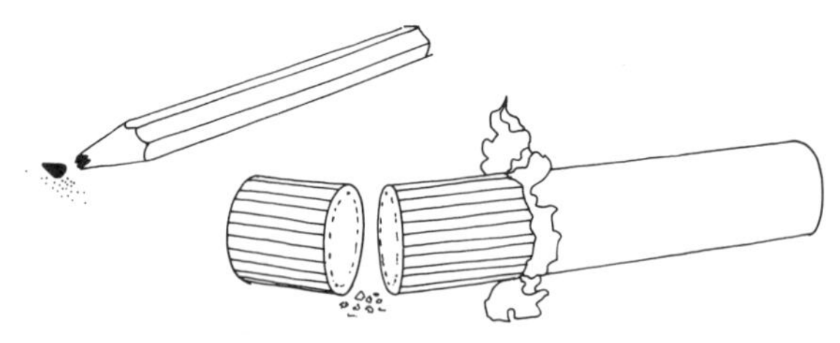

Which is tougher, sheet glass or new putty? The answer should be obvious but is acrylic sheet tougher or more brittle than polystyrene sheet? The designer of a baby's feeder cup would need to know.

Malleability

Some materials are beaten or rolled into shape. Copper or brass bowls can be shaped this way and these materials are said to be malleable.

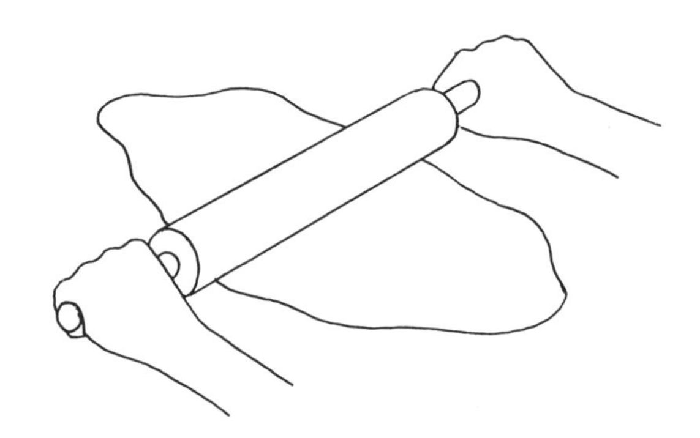

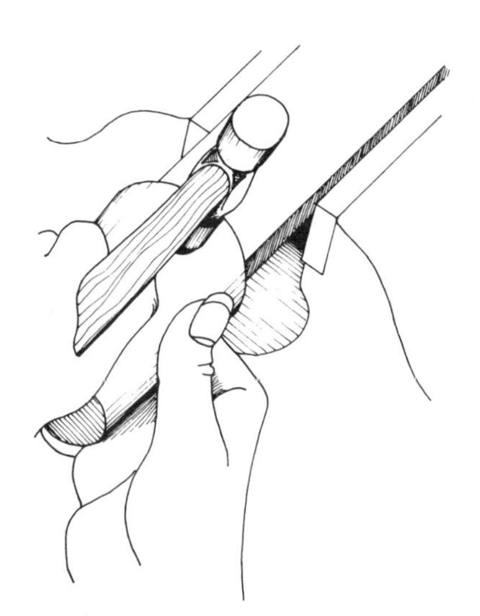

Copper bowl being hammered

Ductility

Can the material be stretched into a long shape without breaking? Copper and chewing gum can be pulled out in this way and therefore have high ductility.

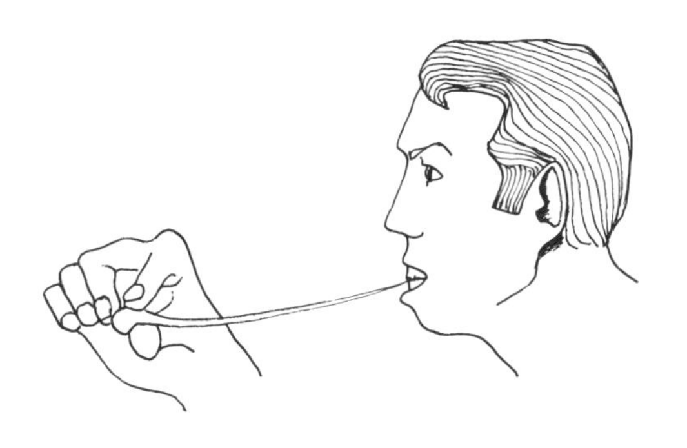

The opposite is low ductility. A material such as lead has low ductility. Any material such as aluminium that can be made into wires is very ductile. A material which is non-ductile is fired clay.

Elasticity and plasticity

How 'springy' is the material? Does it return to its original shape like an elastic band or a foam cushion?

The opposite to elasticity is plasticity. We say that a material that stretches but does not return to its original shape is plastic. Wet clay and putty are plastic.

Compare what happens to a paper bag when it is filled with air and then slapped to what happens when a balloon is partly filled with air and slapped. The section on Energy describes a method for comparing the stretching of elastic bands. See page 51.

Thermal conductivity

Materials which pass on (transmit) heat quickly are called good thermal conductors. Copper and aluminium are good thermal conductors. The opposite to a thermal conductor is a thermal insulator. Woollen fabric and expanded polystyrene are thermal insulators because they trap a lot of air. Air is another good insulator.

Heat-sensitive paper changes colour as it gets hotter. Place samples of steel, copper, wood and a plastics material on a piece of heat-sensitive paper so that they overhang the edge and can be heated gently with a bunsen burner. It is important that the samples have the same cross-sectional area—why? What do you notice about the colour changes. Do not overheat the wood and plastics samples. Can you devise a test for thermal conductivity with the same samples but using candle wax and small ball-bearings to indicate how quickly heat is transmitted?

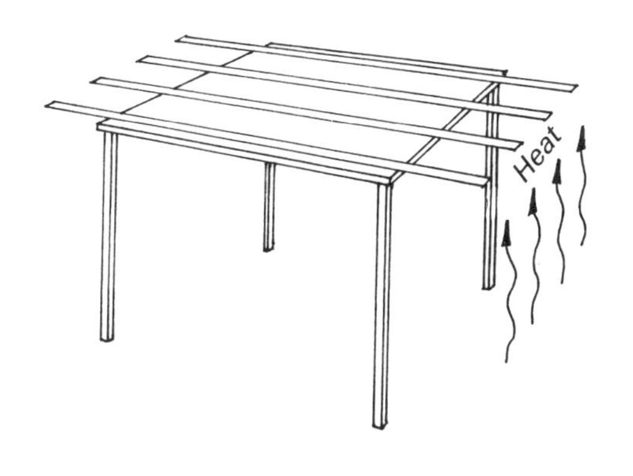

Electrical conductivity

Electrical conductors will transmit electricity easily. All metals are good electrical conductors. The opposite to electrical conductors are electrical insulators like plastics materials and dry cotton.

Make a tester similar to the one illustrated and test as many different materials as possible by holding samples of each between the crocodile clips one at a time. If the bulb lights then the material being tested is an electrical conductor.

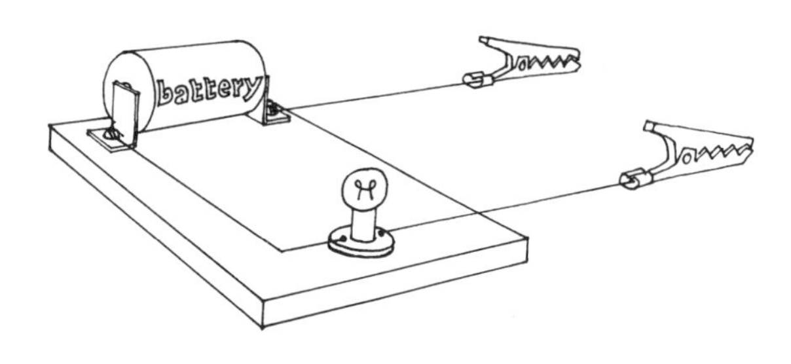

Solubility

Is it possible to dissolve (or melt) the material so that it becomes liquid? Coffee powder mixed with water becomes a liquid. Pebbles would not dissolve in water and are insoluble. A liquid which causes things to melt without the use of heat is a solvent. Petrol causes some plastics materials to dissolve. The fumes given off by some solvents are very dangerous as they can easily catch fire or become addictive. Gloss paint and nail varnish become hard as their solvents evaporate.

COMPARING MATERIALS

Copy the list of properties given below and using suitable materials, fill in each line in the same way as 'HARD' to 'SOFT'. You will find examples of materials in the definition given above but think of examples of your own as well.

HARD	diamond steel wood clay	SOFT
STRONG (tension)		WEAK
STRONG (compression)		COLLAPSIBLE
TOUGH		BRITTLE
HEAVY		LIGHT
CONDUCTOR (heat)		INSULATOR
CONDUCTOR (electricity)		INSULATOR
ELASTIC		PLASTIC

Cost

How much a material costs depends on where it comes from and how it has to be treated before it can be used. Materials such as gold and platinum are expensive because there are small amounts in the world. The opposite to expensive is cheap. Water is quite cheap in countries where there is a lot of it. We use materials that give use the properties that we want at the lowest possible price.

COMPARING COSTS

If you had to decorate a wall at home which do you think is the cheapest way? Try to find out the cost to cover one square metre of wall using each of the following: wallpaper, emulsion paint, gloss paint, ceramic tiles and polystyrene tiles.

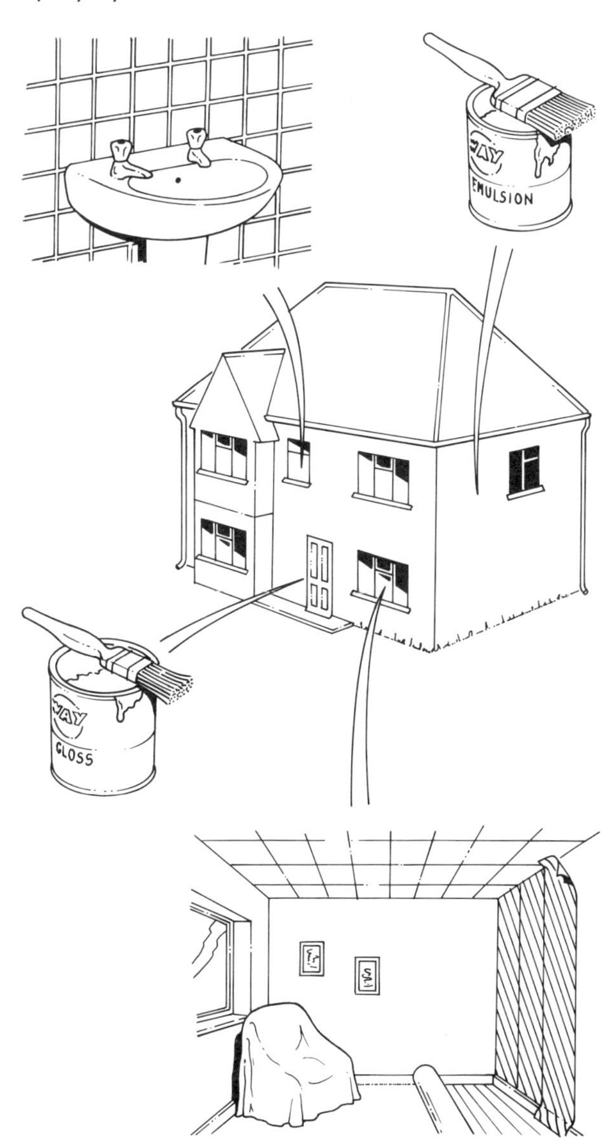

The best way of learning about materials is by using them. The materials most used in CDT are wood, metals and plastics. Here are descriptions of these which will help you to understand where they come from, how they are processed (made fit for use), which special properties they have and how they are formed.

WOOD

Things which are made from wood were once part of a growing tree.

Plants are able to use sunlight to help them grow and so produce stems and leaves. The main stem is called the trunk and people use tree trunks and thick branches in many different ways.

While the tree is growing it is full of watery sap which carries food around the tree. So that we can use it we must season the wood (dry it out). Tree trunks are sawn into thick boards and then stacked so that air can pass between them.

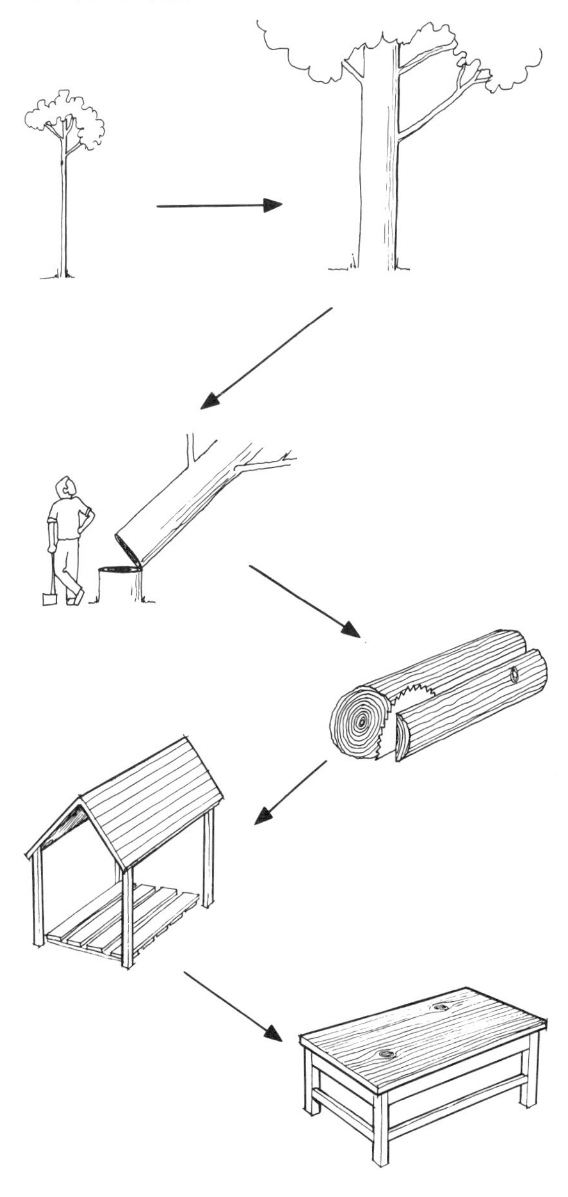

When a tree is cut down, the wood still has a lot of water in it and may twist, crack or shrink. Even when wood is properly seasoned it must be allowed to expand or contract (get bigger or smaller) as it can still take in or lose water.

One of the best things about wood is that because it comes from a plant we can grow more trees to get more wood in the future—it is therefore a 'renewable' material.

Living trees give out oxygen which animals need to breathe so that is another advantage of growing trees.

The most common woods are non toxic (not poisonous). Dry wood does not conduct electricity.

Wood will burn but can be a good thermal insulator. Check that you know what a thermal insulator is (see page 25).

Wood can be sliced into thin sheets which can then be made into plywood. Sawdust, shavings and flakes of wood can be glued together and pressed into large sheets of chipboard.

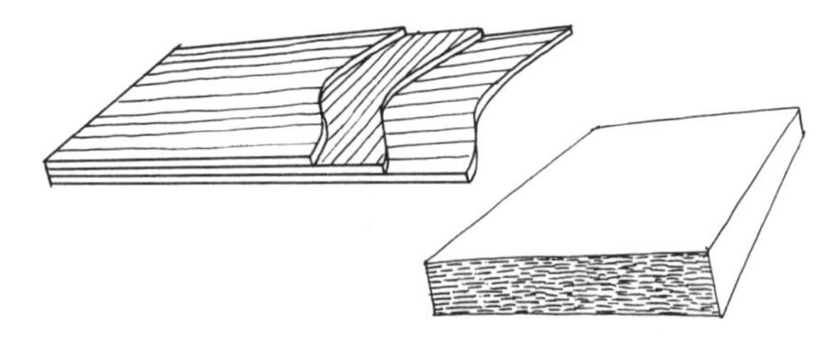

Draw a number of different things which have wooden parts and indicate how the wood has been protected. The most usual ways of protecting wood are as follows: paint, varnish, creosote and melamine (a type of plastic) facing.

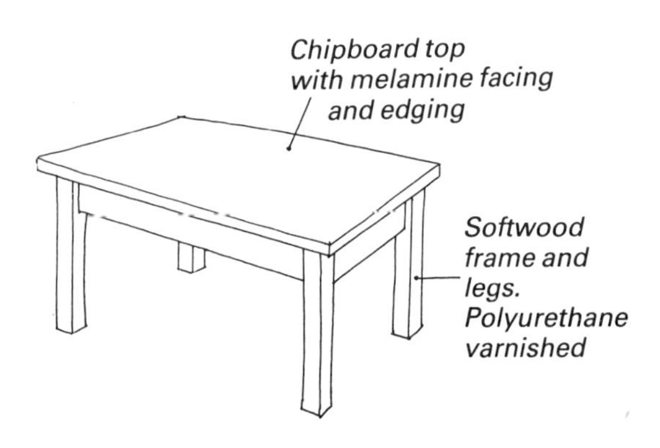

Kitchen table

There are two main kinds of tree: hardwoods and softwoods. The differences between them have nothing to do with hardness (described on page 23) and can only be seen under a microscope.

Examples of softwoods are pine, fir and cedar. Examples of hardwoods are oak, ash and mahogany.

Balsa wood is a hardwood too although it is very soft and can easily be marked by scratching.

Compare the appearance of a hardwood and a softwood by using a sample of each measuring 60 mm long by 30 mm square. Prepare each sample by rubbing one end and one side on a piece of fine glasspaper.

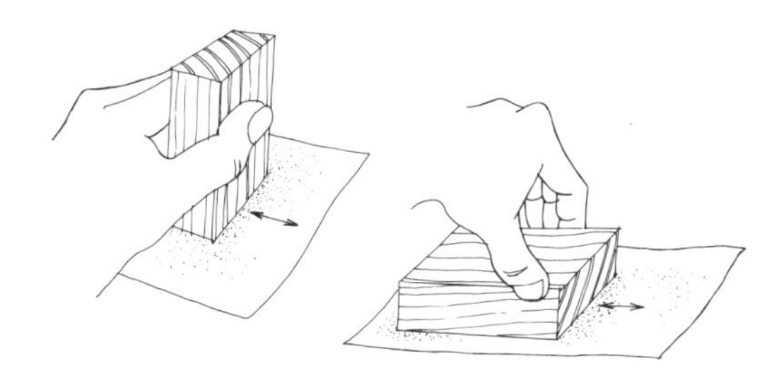

Rub the samples on the glasspaper until you can no longer see saw marks. Look at the end of each sample. What differences can you see? The growth rings should be clearly visible on the end grain of the softwood but you may need a hand lens (magnifying glass) to see them on the hardwood sample. Softwoods grow more quickly than hardwoods. What differences can you find when you compare the side (or long) grain of the two samples?

IMPORTANT FACTS ABOUT WOOD

It has a distinctive grain.
Dry wood is warm to the touch.

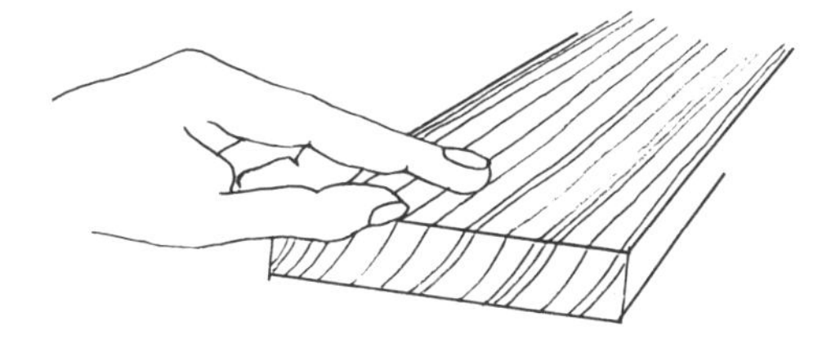

Most wood will float on water.

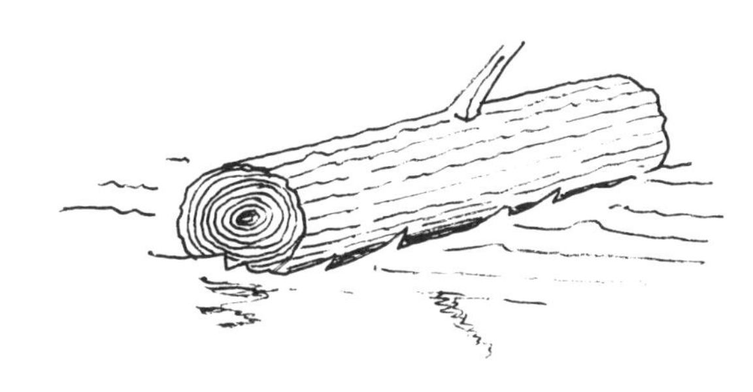

Wood can be easily shaped using hand tools. When making things with wood we need to know which way the grain runs. Short grain is weak.

Cut out the shape shown below. Use softwood that is 6 mm thick. Cut out the shape again but make the grain run at right angles to that of the first example.

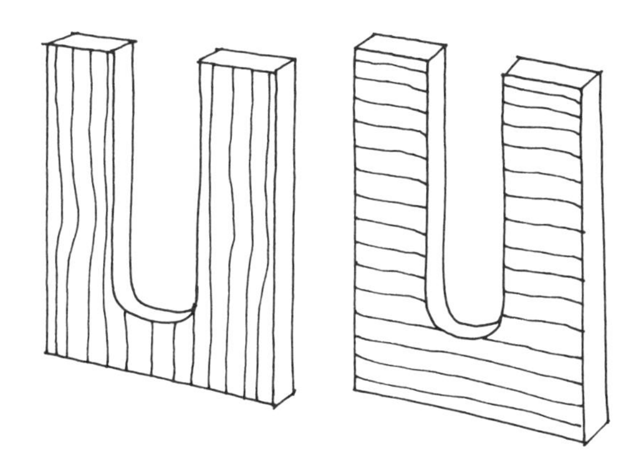

You may have found that they broke whilst you were cutting them. Where did they break? If they did not break try to break them. The wood usually breaks where the grain is short.

PAPER

Making paper

Most of the paper that is used for newspapers, cardboard and drawing paper is made from wood pulp. The pulp is pressed into sheets by rollers. The thickness of paper is measured by its weight per square metre. Drawing paper may be 120 g/m^2 (grammes per square metre) and cardboard may be 350 g/m^2. The properties of paper depend mainly on its weight. Cardboard would have a greater tensile strength than cartridge paper. It is important to consider the paper's wet strength—whether it is easily affected by water.

Using Paper

Paper can be written on, printed on and can be embossed (have patterns pressed onto it like some wallpaper).

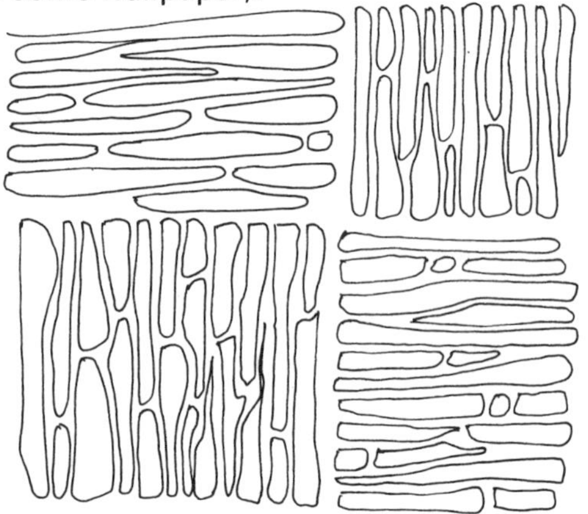

It can be easily formed by hand tools and can be glued, stapled, clipped and taped. It is non-toxic (not poisonous). Dry paper is an electrical insulator. Check that you know what an insulator is from page 25.

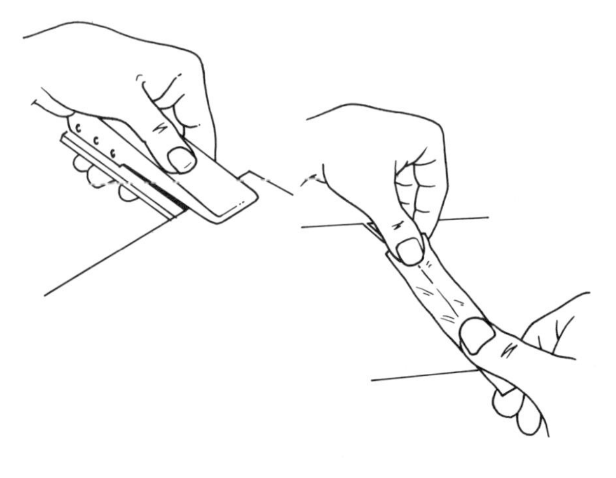

Paper is also a good thermal insulator. It is an inexpensive modelling material. It gives cheap way of passing on information through newspapers, textbooks, drawings, notes and computer print outs.

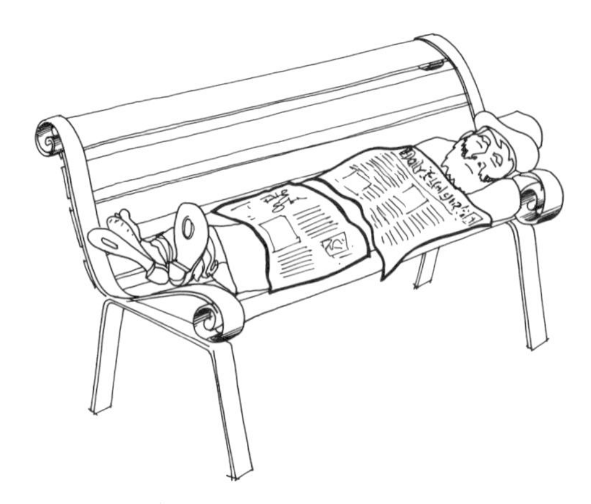

How is the newspaper being used here?

Paper can be made into strong and complicated shapes by mixing it with glue, pressing it onto a mould and allowing it to dry. This is known as papier mâché. Paper, like wood, can be attacked by insects and fungi and is therefore known as biodegradable. Packages and cartons made from paper alone can be broken down by natural means and are easier to dispose of than plastic containers.

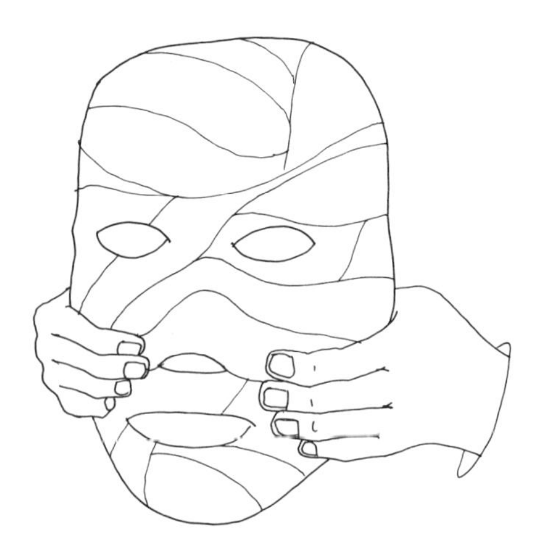

The size of a sheet of paper is described by an international code. Sizes vary from A0 (a large size) to A7. The most common sizes are A2, A3, A4 and A5.

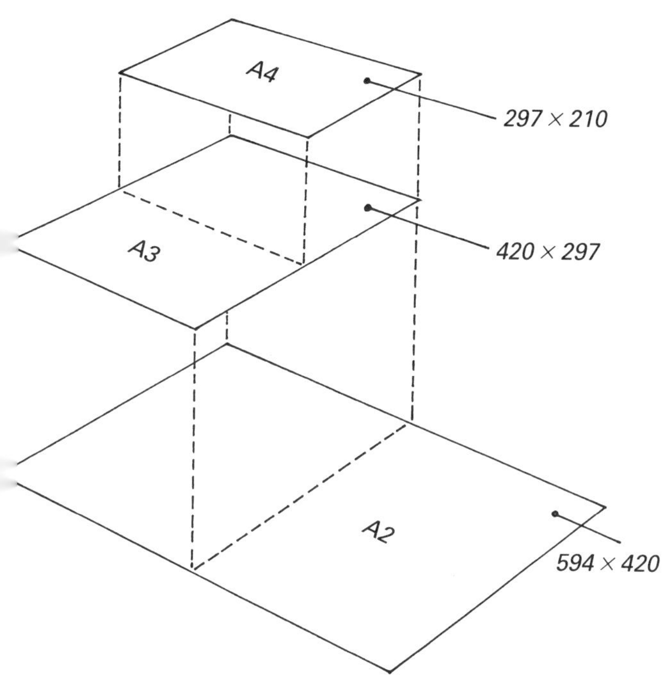

Investigating paper structures

Divide a sheet of A4 drawing paper (about 110 g/m^2) by folding into four equal parts.

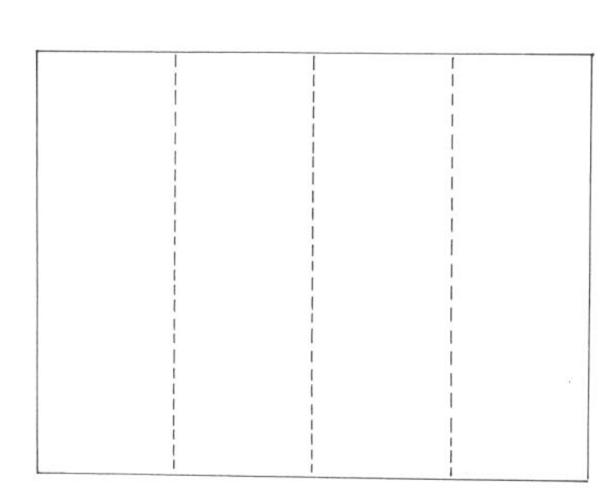

Cut along each fold and roll each section around a pencil to make a tube 210 mm long. Glue the end of the rolls and remove the pencil.

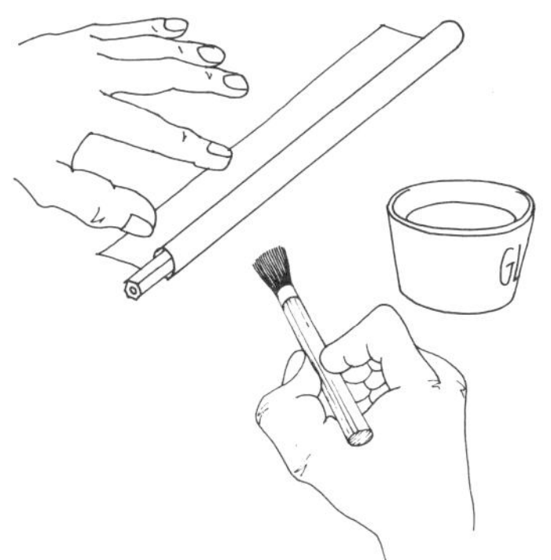

When the glue is dry support the ends of a tube and pull gently on a force meter at the tube's centre. (The meter will need to read in 0.05 N). Note the reading when the tube starts to crumple.

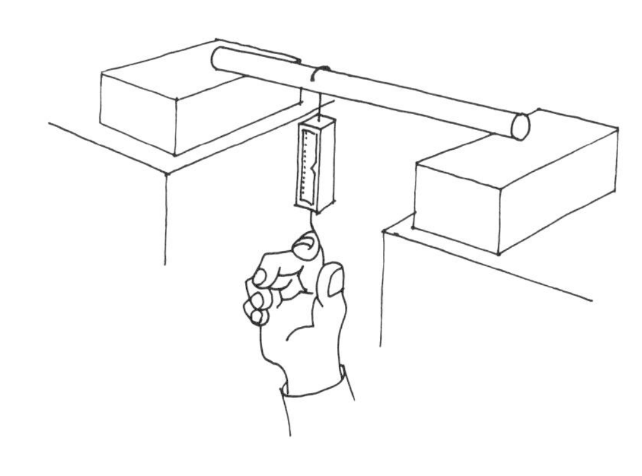

Repeat the test with the remaining three tubes lightly glued together.

How much force is needed to crumple the three tubes compared with the force to bend one? Why is it important not to use too much glue?

Projects using paper (1)

Would it be impossible to balance a house brick on the edge of a sheet of A4 drawing paper? How could you cut and join a single sheet of paper so that a brick is supported at a distance of at least 75 mm above the table top? Be careful when placing the brick on top of your structure(s) as it may suddenly topple over.

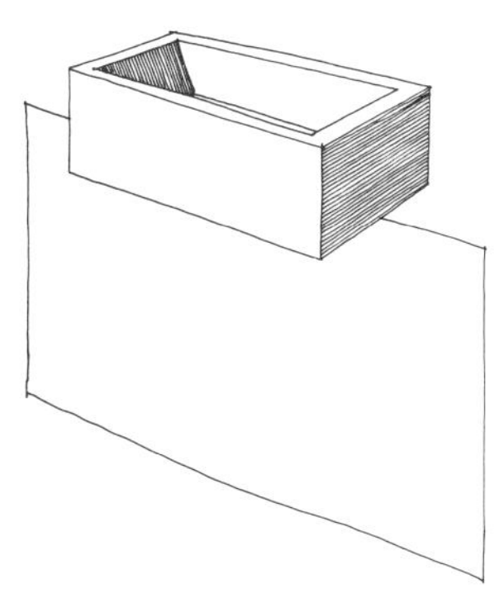

Can you balance a brick on the edge of a piece of paper?

Try to support the brick in at least three different places. When the paper tubes are loaded in this way, what word from page 26 can be used to describe their strength?

Projects using paper (2)

By using only three sheets of A4 drawing paper make the tallest possible structure that will support a tenpence coin (or 10 g mass) at the top. You do not have to use cylindrical (round tube) shapes.

METALS

Metals containing iron are called ferrous metals. All other metals, such as aluminium, lead, tin and copper are called non-ferrous. Some metals can be mixed together to produce another metal which is different from those used to make the mixture. This new metal is called an alloy. An example of this is where lead and tin are mixed together to make Solder. Lead and tin are pure metals but solder is an alloy.

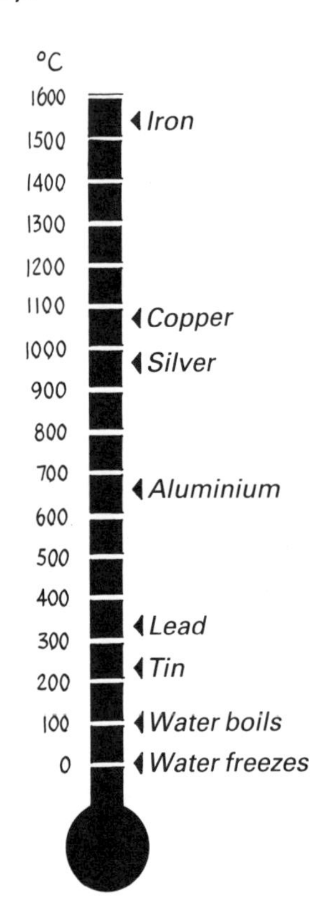

Temperatures at which some metals melt

The temperatures at which pure metals melt vary over a wide range. Alloys have a different melting point to the metals from which they are made. They can also have properties different from the original metals.

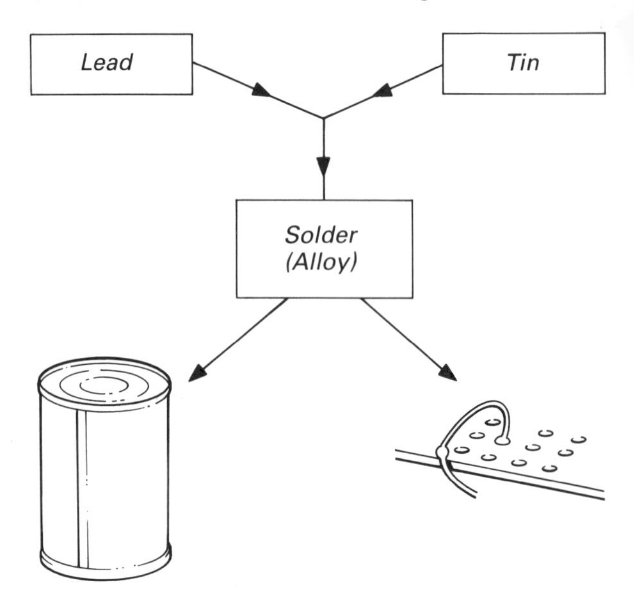

Solder melts at quite a low temperature and is used to join electrical wires together or make seams in things like 'tin' cans. Changing the quantities of lead or tin, changes the melting point of solder.

Joining metals investigation

To join two pieces of steel using solder, take two strips of thin mild steel sheet measuring 50 mm × 25 mm × 2 mm. Gauge and clean the area to be joined by using emery cloth. Paint active flux (see the Service Section) onto the area to be joined and overlap the two pieces. Place small pieces of tinmans solder along the line of the joint and heat with a blowtorch. When the solder melts and runs between the pieces of metal stop heating and allow to cool.

Use the hottest part of the flame, which is the tip

Repeat the above using two more pieces of mild steel sheet but using a brazing flux. Place small pieces of brazing spelter next to the joint.

Does brazing spelter (which is mostly brass) need more or less heat to melt and make the joint than solder (mostly lead)?

Which method makes the stronger joint? How could you test the joints to find out?

Use the protective clothing and equipment provided by the school. Use tongs if you have to move the metal whilst it is hot.

By mixing molten iron and carbon, steel is produced. Steel is much harder than iron. By mixing copper and zinc together brass is produced. Brass is harder and more elastic than either copper or zinc.

Metals conduct electricity. Ferrous metals attract magnets. Ferrous metals which get wet and are exposed to air will oxidise (rust). Non-ferrous metals also oxidise but not usually enough to weaken them.

Aluminium oxidises in air, but the oxide film is clear and protects the metal underneath it from further oxidation.

Steel car bodies rust if not protected

Small metal parts can be glued together but the joint must be perfectly clean before doing so.

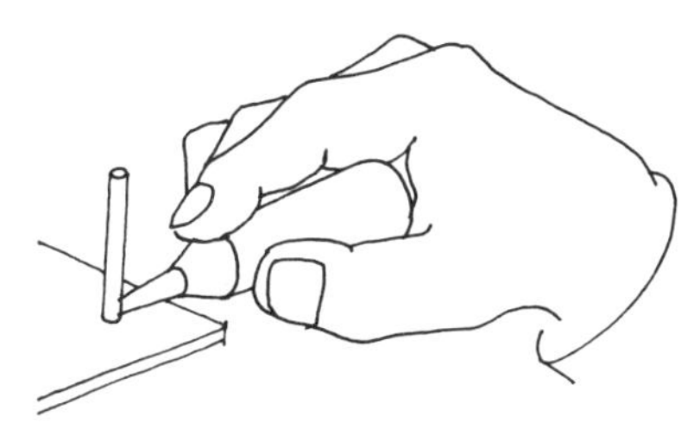

Steel parts can be joined by welding—melting the edges of the metal together so that, when cool, they become as one.

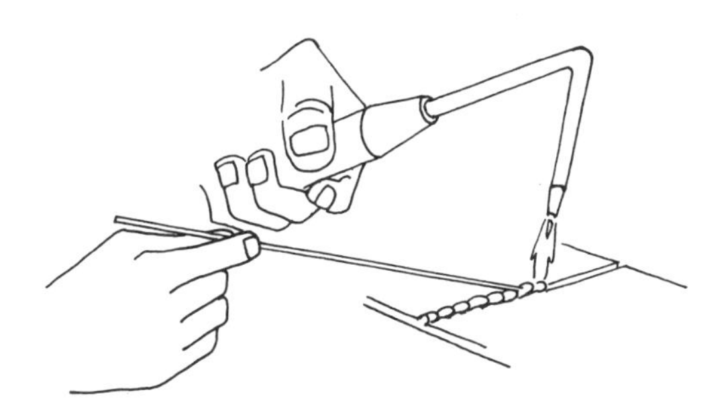

Some metals, such as lead, are toxic (poisonous).
What does this sign mean?
Where might you see it?

Some types of steel which contain about 1% carbon can be made very hard by heating to bright red and quenching (cooling) quickly in cold water. The metal may then need to be heated again so that it becomes tempered (tough) as well as hard.

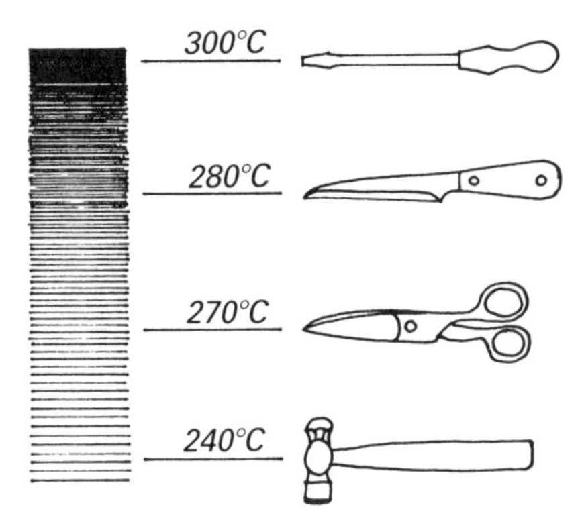

The temperatures at which different tools are tempered

Investigating metals

Take a strip of aluminium alloy, work-harden it by hammering and then try to bend it. Coat with soap and heat gently until it blackens. Clean up the metal and try to bend again.

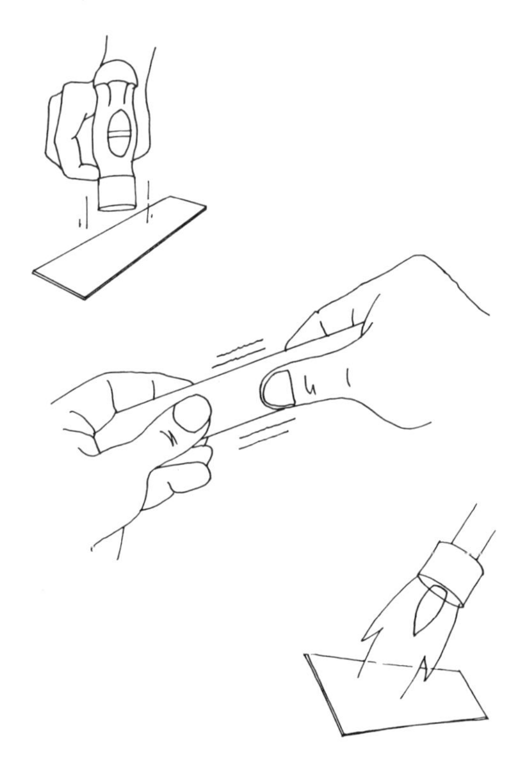

Molten metal may be allowed to cool slightly as it runs along channels so that it ceases to be a liquid but stays soft. The metal can be rolled into sheet (like pastry) or squashed into slabs or dragged into long thin rods or wires. As the metal cools it becomes less easy to form so it has to be shaped quickly.

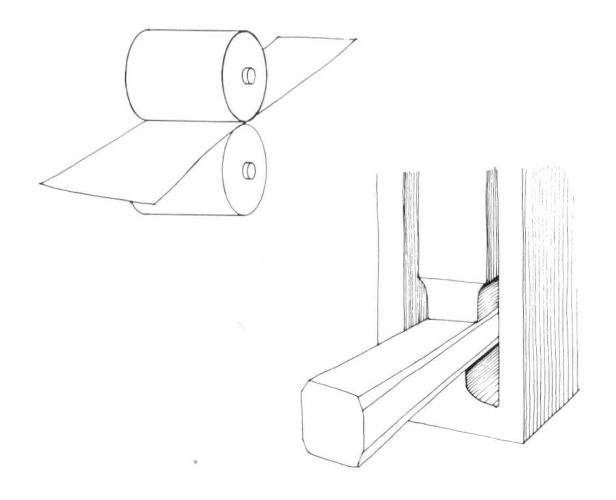

Most metals require very high temperatures to melt them. However once the metal has become molten it can be poured into a mould and allowed to cool (like a jelly). When the metal is cool it will have taken the shape of the mould. This is called casting. Brass ornaments and iron benches are made by casting.

Make a list of 10 objects which are made by casting. Why is casting such a useful way of shaping metal?

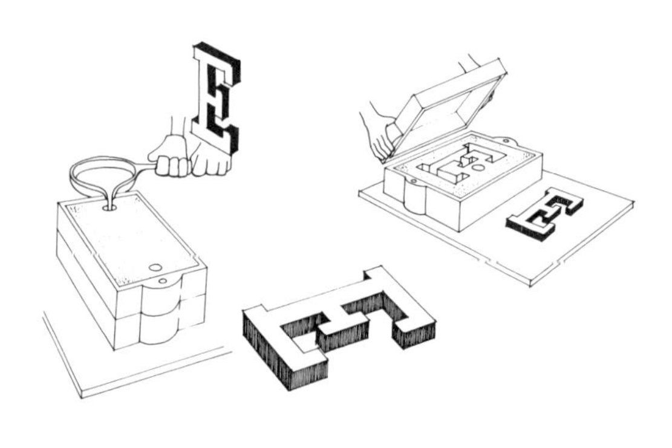

Sand casting

Projects using metal (1)

The diagram shows a clinometer which can help you to work out the height of a tall object.

The clinometer is used by looking through the sights to the top of the object and allowing the pointer to swing until it points straight down. A reading can be taken from the protractor scale which is used in the calculation:

$$\text{height} = \frac{\text{distance}}{\text{the tangent of the angle}}$$

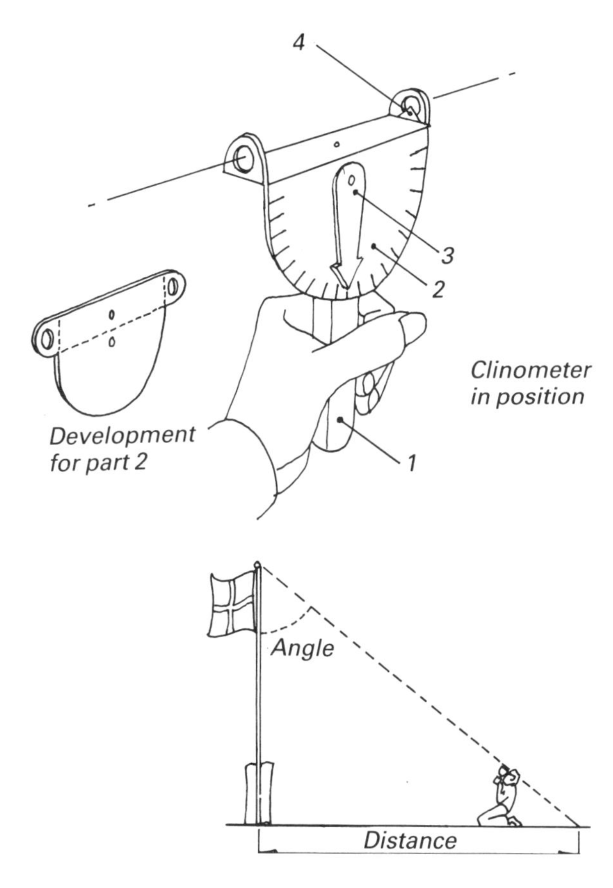

Using the clinometer

The sizes used can be based on the size of the protractor scale and the size of your hand. Make a paper protractor scale so that the distance between the sights is about 100 mm. What materials do you think would be best for the parts numbered 1–4? How many ways can you think of for joining parts 1 and 2 together? What kind of fit must there be between parts one and three? (Should the joint be tight or loose?)

Projects using metal (2)

Decide which material would be best to make the parts for the balancer shown in the illustration. Parts 1 and 2 must be brazed together and parts 2 and 3 must be soft-soldered together.

The only materials which can be joined in this way are metals. Which two main types of metal are there? You will not want the project to be too expensive.

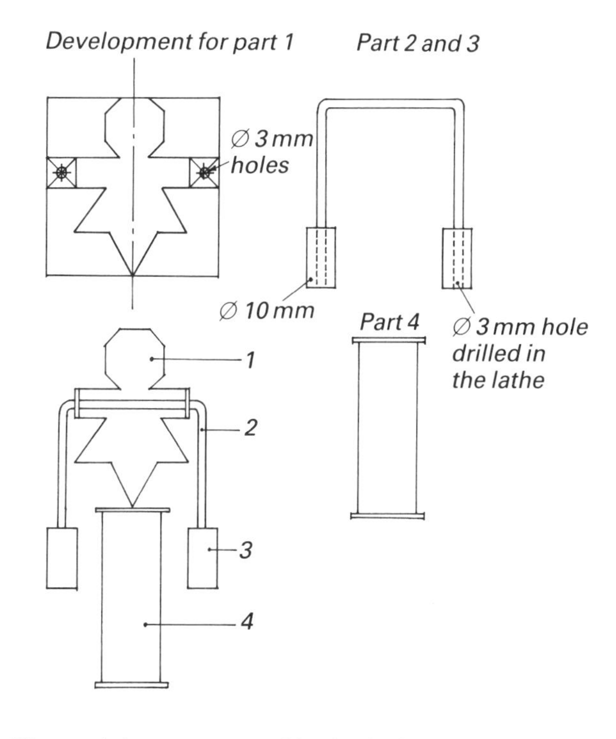

The weights are turned in the lathe and soft-soldered to the rod.

Design a base using a 50 mm length of mild steel tube. Clean the metal using emery cloth and then paint it. Find out the answers to these questions while doing this project:

1. Is mild steel ferrous or non-ferrous? (See page 32)
2. Why do you need to use a centrepunch before drilling a hole? (See the Service Section.)
3. What does a 'flux' do? (See the Service Section.)
4. What is an alloy? Is steel an alloy? (See page 32)
5. Why is it necessary to paint ferrous metals (like mild steel)?
6. Why does the balancer balance?

PLASTICS

Many of the plastics that we use today have been developed during the last fifty years. Most plastics materials are made from oil. Crude oil is brought to the surface through an oil well. The oil is separated into fuels such as petrol and paraffin and also into chemicals from which plastics can be made.

An oil refinery

There are two main types of plastics, thermosetting and thermoplastic. Thermosetting plastics soften and melt when first heated but then set hard permanently (for ever). Things which may get hot such as electrical sockets are usually made from thermosetting plastics. An accurate metal mould (called a compression mould) is filled with a thermosetting plastics powder and heated as great pressure is applied to the two halves of the mould. Once the mould has cooled the compression moulding can be removed.

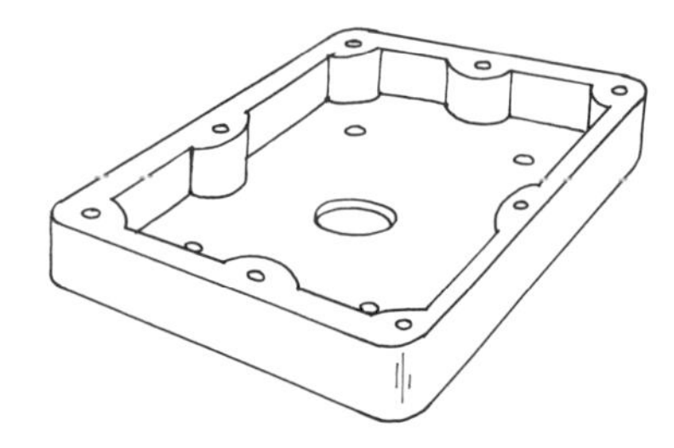

Part of an electric socket made by compression moulding

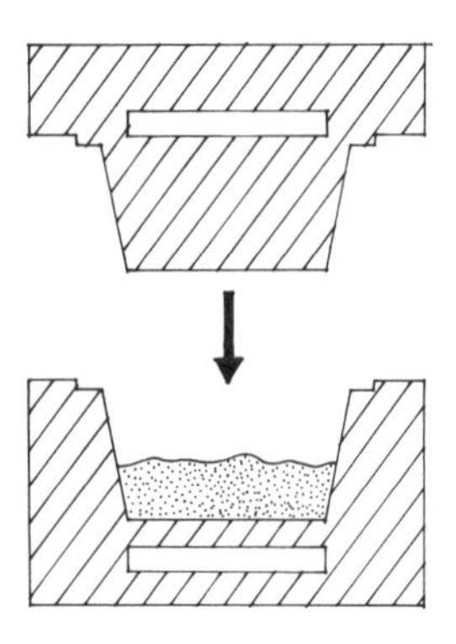

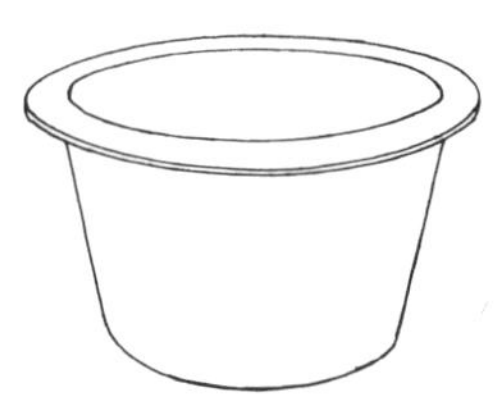

Compression moulding a bowl

Find examples of other plastics items that may get hot in use. Start your list with saucepan handles and soldering iron handles.

Thermoplastics soften and melt but set when cool. Provided that they are not overheated, thermoplastics materials can be softened by reheating many times without damage.

Thermoplastic items can be formed in a variety of ways. These methods include: extruding shapes through a die (like squeezing toothpaste from a tube), injecting molten plastic into a mould as shown below and heating sheets of plastic so that they can be pushed or blown into a new shape before cooling.

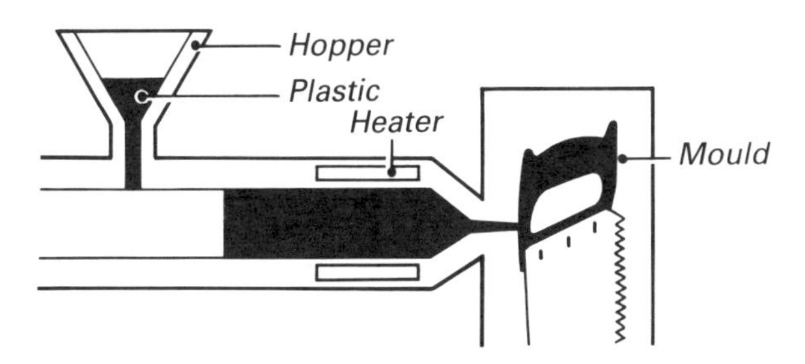

An injection moulding of a saw handle

Why is it important that thermoplastics materials are not used for a fitting which is to hold a 100 watt lightbulb?

An injection moulding machine

Investigating thermoplastics

One way of making a container such as a 'yogurt' pot or margarine tub is for the manufacturer to take a sheet of thermoplastic material, heat it and by using a vacuum forming machine, suck it into a mould. If you heat a tub gently, the thermoplastic material will slowly change its shape back to the original sheet form. This is called the plastic's 'memory'.

See what happens when you put an empty margarine tub into boiling water. Be careful when doing this. Use tongs when placing the tub in the water but above all make sure that you can use the boiling water safely. What do you notice about the speed at which the change takes place? Which parts will change more quickly, thick or thin? Which is the most stretched (thinnest) part of the container? A more rapid change will take place in hotter surroundings—like a domestic oven. Make sure that you have permission to use the oven and place the tub or pot on a metal sheet. Set the temperature control to give a heat of no more than 200°Centigrade. The plastic may stick to the sheet or set on fire if it gets too hot.

Materials such as polystyrene (a thermoplastic) can be formed into sheets and tubes which are suitable for covering food or making bottles. As plastics are waterproof they can be used to keep moisture in and germs out.

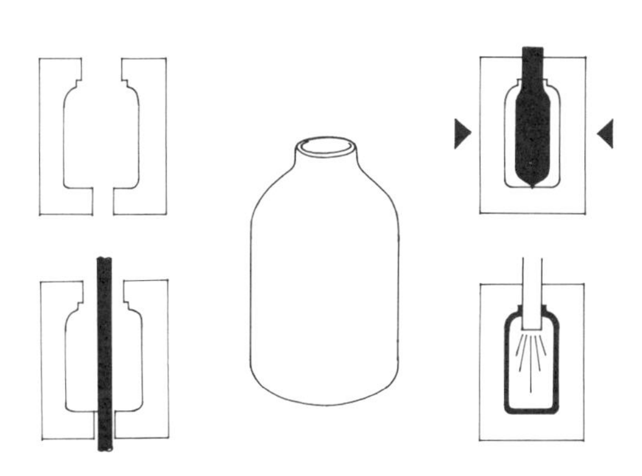

Washing up bowls, sandwich boxes, saw handles, combs and cassette boxes are not intended to get hot and are usually made from materials like polyethene, polypropylene acrylic and polyvinylchloride (PVC).

Plastics materials can be mixed with other plastics or non-plastics to make new materials with different properties. Plastics are man-made, they can be designed to suit particular needs. There are hundreds of different plastics which are known by their scientific as well as their trade names. The scientific name for nylon is polyamide and for polystyrene is polyphenylethene. Nylon is used to make carpets and strong mouldings whilst polystyrene is used for making disposable cups and packaging materials (in its 'frothy', expanded form).

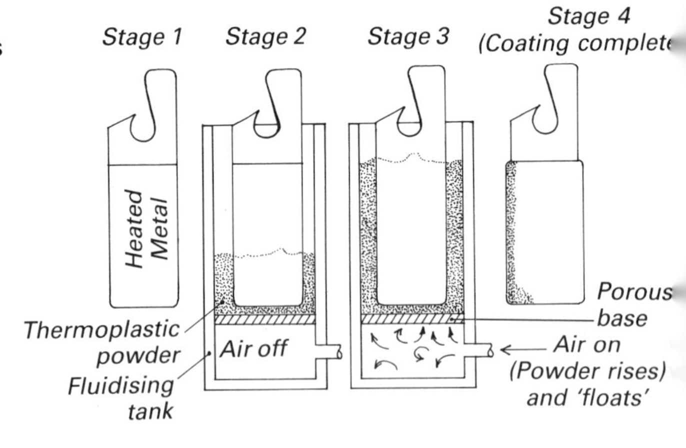

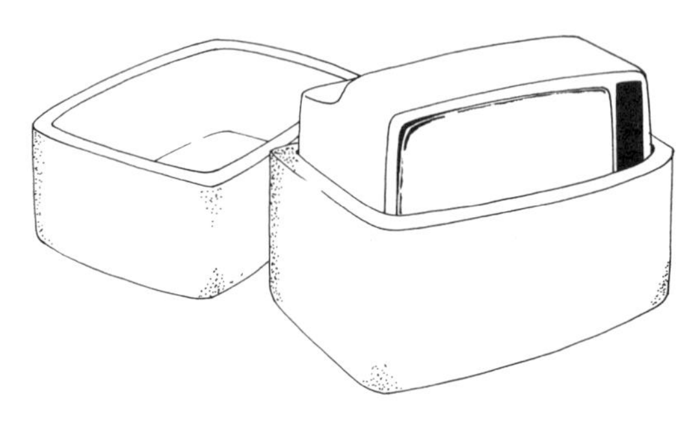

Polystyrene packaging

Bending thermoplastic sheet such as acrylic or PVC can be carried out using a strip heater which heats a narrow strip of the material for straight line bends.

Make sketches of articles you could make by bending thermoplastic sheet.

Plastics materials can be used to cover wood, metal or paper products to provide hard wearing worktops, metal draining racks which do not rust because of their plastic coating and papercups which do not leak because of the thin layer of plastic which covers them.

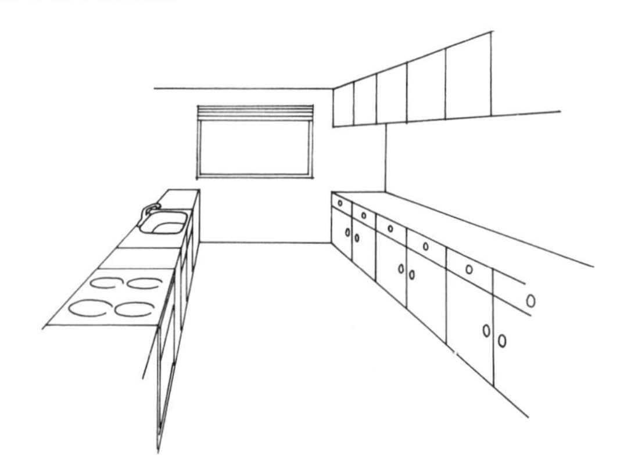

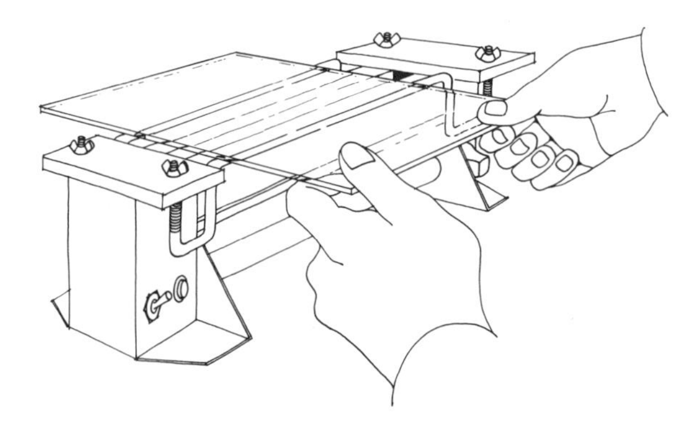

Using a strip heater

Metal shapes can be heated to about 200°Centigrade and dipped into a thermoplastics powder such as polythene. As the powder melts, it forms a skin.

Plastics can be shaped using hand and machine tools such as scissors craft knives, saws, files, drills and lathes. Plastics can be joined by the use of adhesives (glues) and thin sheet or film can be seamed by heat sealing machines.

Using plastics (1)

Compare wood, metal and plastics materials by making the coat hook using the pattern as a guide.

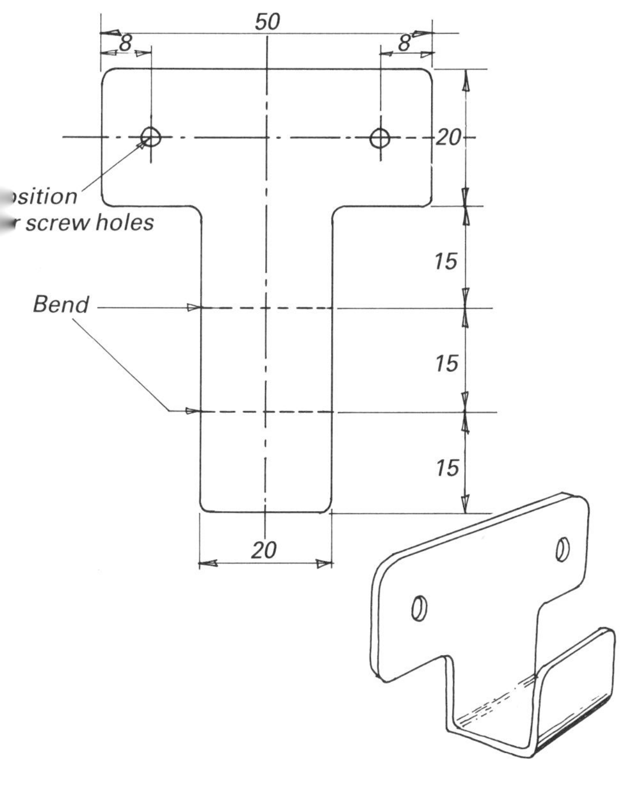

Use 3 mm thick plastics sheets (PVC, acrylic or polypropylene).
Use 3 mm thick plywood.
Use 1 mm thick mild steel.
Round off and polish the edges of the metal and plastics before bending.

When heating the plastics sheet the strip heater will make the material very hot at the bends. A wooden holder like the one shown in the illustration will help you to handle it and will not mark the plastic.

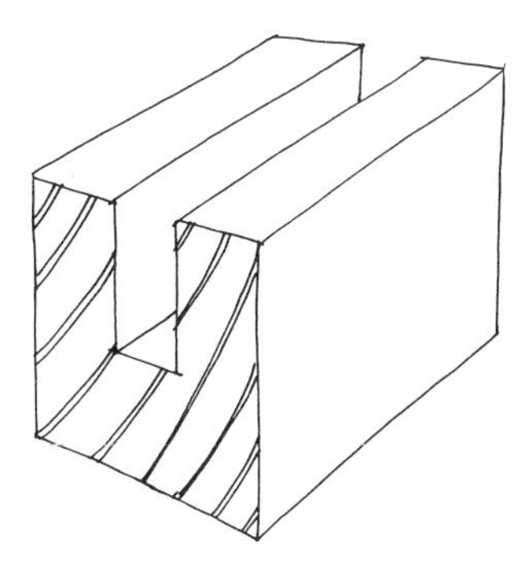

Which property of wood are you making use of? (See page 25) Use panel pins and glue to join the plywood. Is this easy on such a small piece of work? Are there better ways of making such a hook using wood?

How difficult is it to bend the metal version in the vice? What would you have to do to the metal version to stop it rusting? What surface treatment does the plastics hook need? Why would PVC be better than acrylic for making the hook?

Using Plastics (2)

By using a mould similar to the one shown you can make a container from thermoplastic sheet.

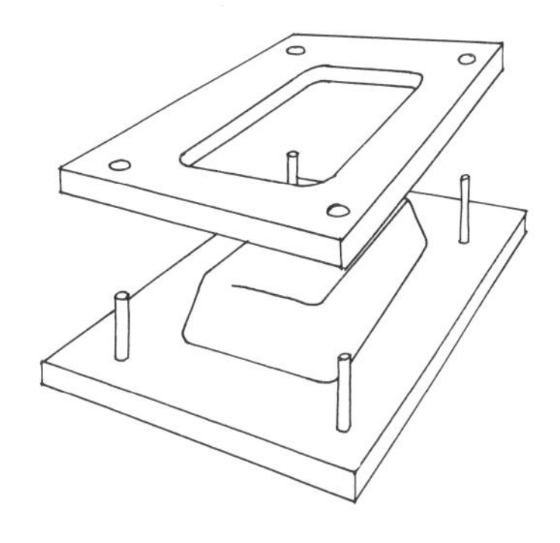

Heat the sheet for about 20 minutes in an oven set to no more than 200°C. Handle the hot plastic with care by using pliers and making sure that you do not burn yourself on the plastic or the oven. Use the safety equipment provided by the School.

Make sure that the sheet drapes evenly over the raised part of the mould before clamping down the part with the hole in it. After 10 minutes separate the mould and remove the plastic. You will need to trim and polish the edges.

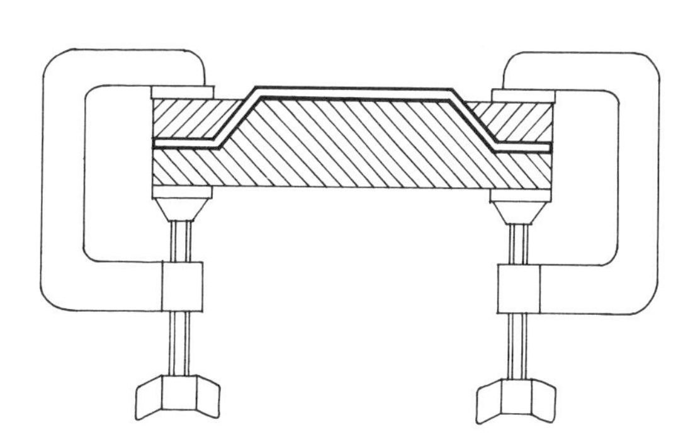

The container can be used for fishing tackle, electronic projects, sewing kit or any other purpose. Can you design and make a lid or sliding cover to fit the container? By making suitable tracks the container could be made to fit under a shelf to slide out like a small drawer. Use a clear acrylic sheet if you want to see the contents but PVC, high impact polystyrene or polypropylene are all suitable for this project.

How good was your tray? What could you do to improve the quality of the moulding?

How easy would it be to make a metal container by using the same method?

Are there any other materials that are transparent (see-through) that could be used for the container?

Materials project

Use the diagrams to help you make a puzzle. You have to put the parts together in a special way to make them interlock (hold together). You could use almost any material but some would not be suitable. Choose a material which could easily be cut by hand and which is cheap enough for you to use a piece which measures 405 × 45 × 15 mm. The material will need to be rigid and tough. Which material will you choose? Why would you choose it?

Mark the material as shown and cut out the shaded parts before the 405 mm length is cut into three. Take care to cut on the waste side of your lines and then use a file to make the cut accurate.

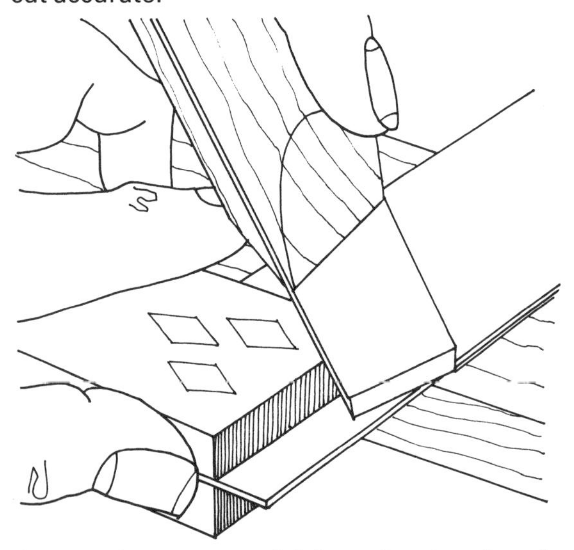

Squeeze the try-square tightly against the edge of the wood

Try drawing the 'exploded view' of the parts on an isometric grid like the one shown on this page.

You can change the dimensions of the blocks but you must keep the same proportions or your puzzle will not fit together.

Try painting each block a different colour.

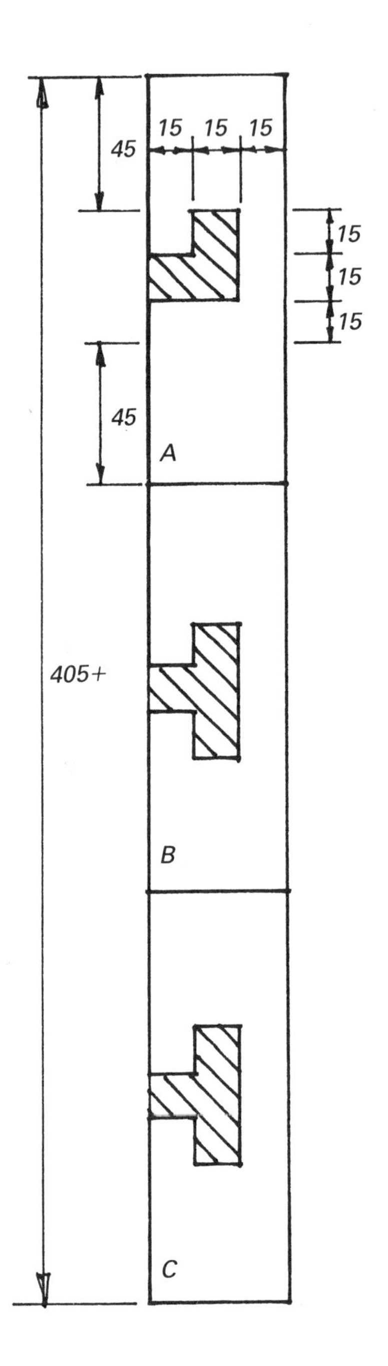

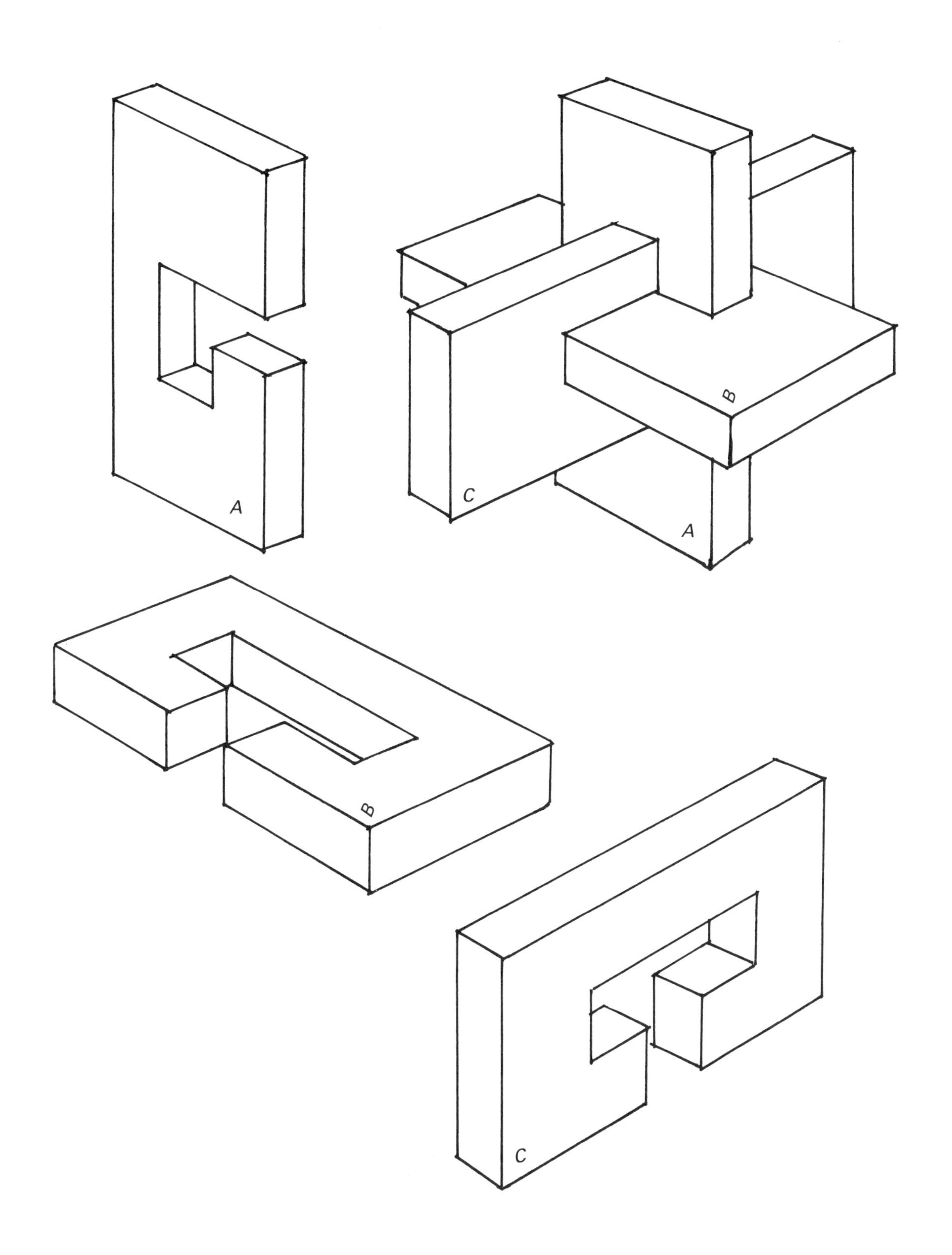
A
C
B
A
B
C

Structures

Materials can be formed into a variety of shapes, for instance sheets, tubes and angles. The shape of the end of a material is called its cross-section. Knowing the cross-section of a material is very important. A tube made from a material may be very strong and would save material as well as being lighter than a solid shape.

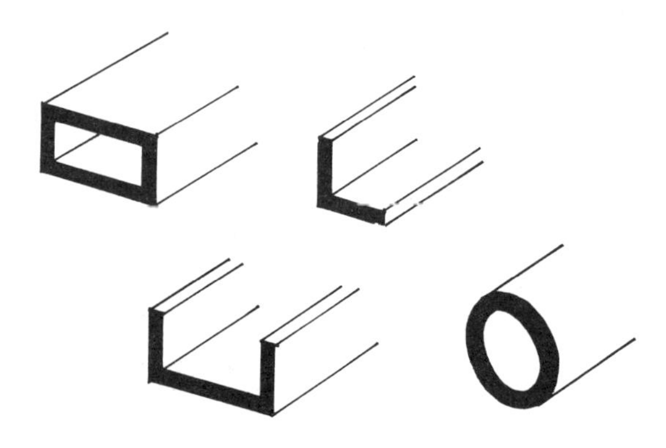

Frameworks are structures made from the shapes shown in the illustration. Examples of these can be found in builder's scaffolding, chair frames and bicycle frames. All of these need to be strong but light enough to be easily moved.

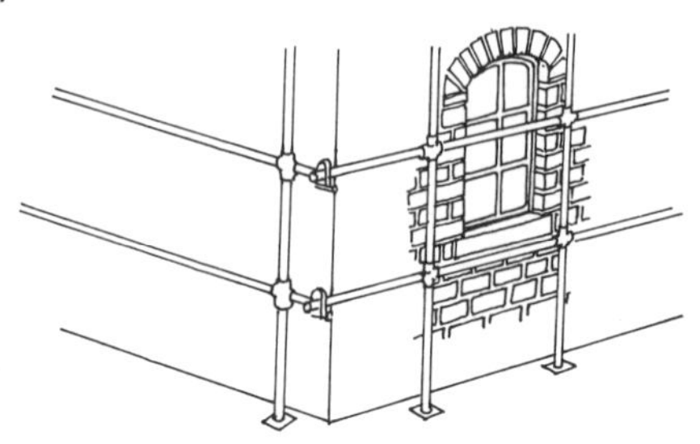

A framework must not only be light but also rigid. Rigid means that the parts of the framework will not move about once they are in place. A way of showing this is to make a framework from the flat strips of a 'Meccano' set. Make a frame as shown in the illustration without tightening the nuts and bolts at the corners too much. The shape of the frame will change if one corner is pushed. Using another flat strip join it to the frame as shown in the illustration. This extra strip will make the framework rigid because the structure has a triangle in it.

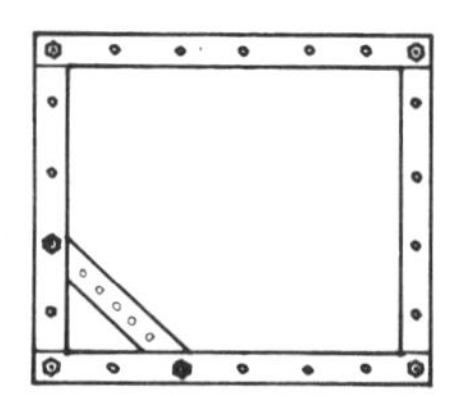

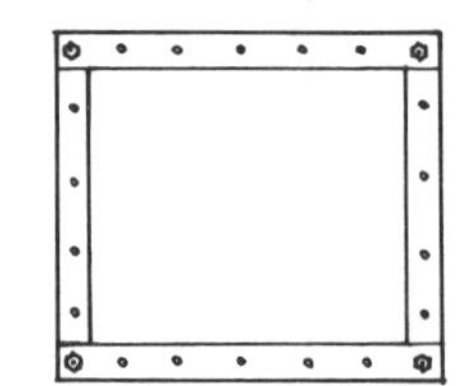

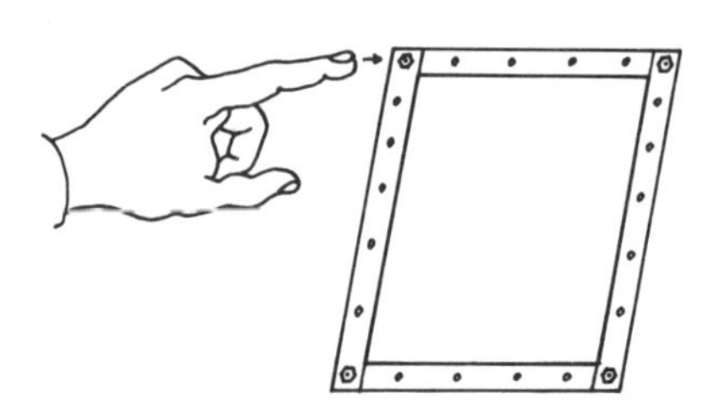

The triangle is a very strong shape and is used wherever framework structures are found, for example, in girder bridges.

Bridges can be made in other materials—older bridges are made of stone and usually have an arched shape.

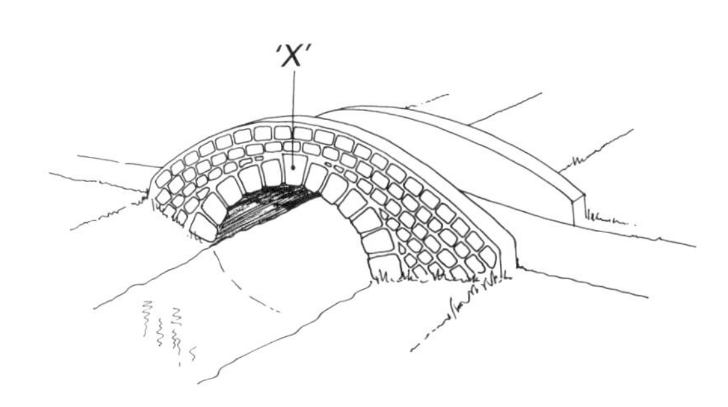

Why are stone-built bridges arched? What would happen if the key-stone, X, was removed?

Large modern road bridges may be of the suspension design which means that the roadway is carried by wire cables which are supported by towers. The bridges over motorways are often made of concrete and are called beam bridges.

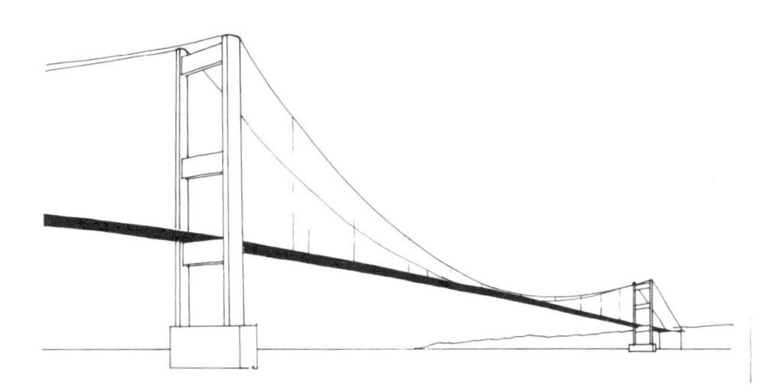

Investigating structures

In all bridges some parts are in tension (being stretched) and some parts are in compression (being squashed).

Copy the diagram of the tightrope walker and use arrows to show tension and compression in the rope and also the framework. A part which is in tension can be shown by arrows which look like this: ——→ ←——. Parts which are in compression can be shown by arrows which look like this ←—— ——→ .
An example of how this system works is shown on the diagram of a rope bridge.

Parts which are in tension are called ties. Parts which are in compression are called struts.

Structures project (1)

Design and make a bridge to span a gap of 230 mm to carry the largest possible load at its centre.

Try to use some of the ideas that you find in your research. Draw at least four different designs. Select the best idea and draw a full-size diagram of one side only. Do not try to draw the whole bridge as this may prove too complicated. To make the structure you may use any of the following: 2.5 mm square softwood; balsa wood; card or thread. Joints can be made using PVA glue and where a number of parts meet a cardboard gusset can be used.

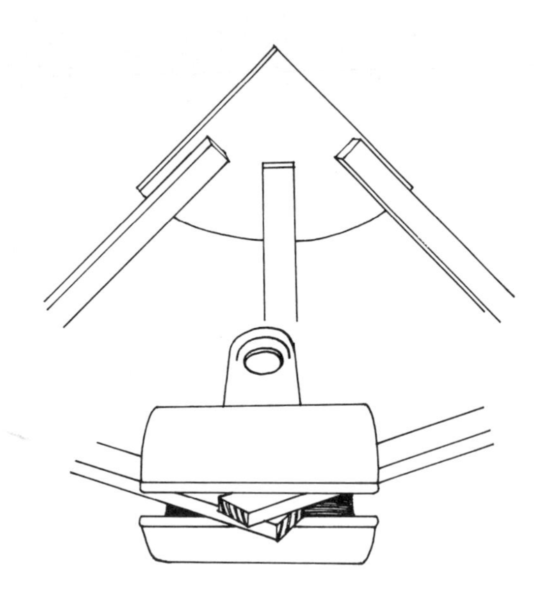

Place the parts over the lines of your diagram and hold the joints together as the glue sets with bulldog clips or drawing board clips. Make two frames in this way and join them with pieces which are 60 mm long. Make sure that there is somewhere for a loading block to rest. When complete, weigh the structure.

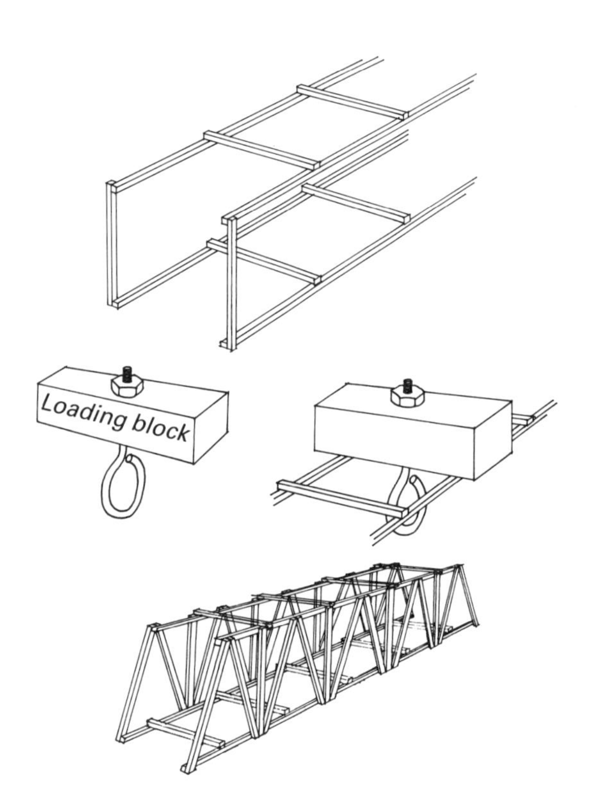

Make a gap to bridge by moving tables close together or make a strong test-rig. Suspend a bucket from the loading block and fill slowly with damp sand. When the bridge starts to fail, weigh the bucket. Beware the bucket may fall suddenly without warning.

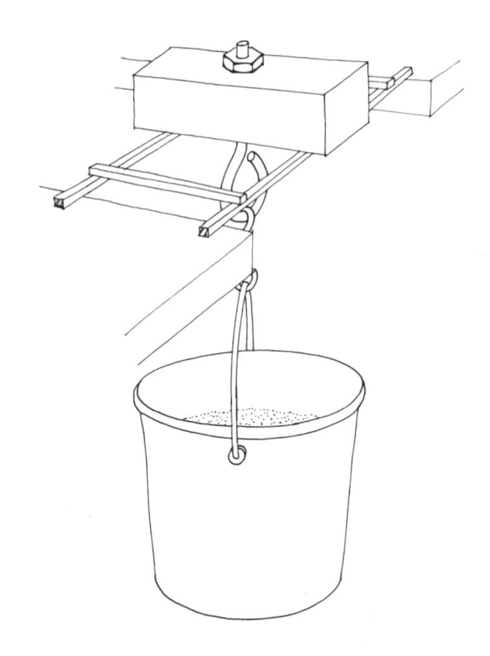

Take the load at which it failed and divide this by the weight of the structure. This will give a load to weight ratio. If the structure weighs 50 g and it fails at 3,500 g (3.5 kg) the load to weight ratio is:

$$\frac{3{,}500}{50} = 70$$

which is 70:1.

Compare the ratio of different designs of structure.

What went wrong with your structure? How could you improve it?

Structures project (2)

Design and build an emergency shelter for two people using the following materials: 12 garden canes each 2 m long, 5 m of polypropylene string or any other other strong string, 5 m^2 of polythene sheet and 15 bricks (or equivalent).

The shelter must resist the wind and weather and be only just large enough for the two people to sit and lie down. Write a description of how you could test the shelter and how you could improve it.

Energy

Energy is needed before anything can move or do work. We work when we breathe, run or think. Almost all of the world's energy comes from the sun.

The Sunraycer, a car powered by the sun's energy

Investigating energy (1)

On a sunny day use a magnifying glass (lens) to focus (aim) the Sun's rays onto a piece of scrap paper. If you hold the lens still the paper will catch fire. Do not do this where the lighted paper will cause damage.

Do not look at the sun through a lens as it will burn your eye. Why should empty bottles not be discarded in hedgerows?

The illustration shows how the sun is connected to other forms of energy.

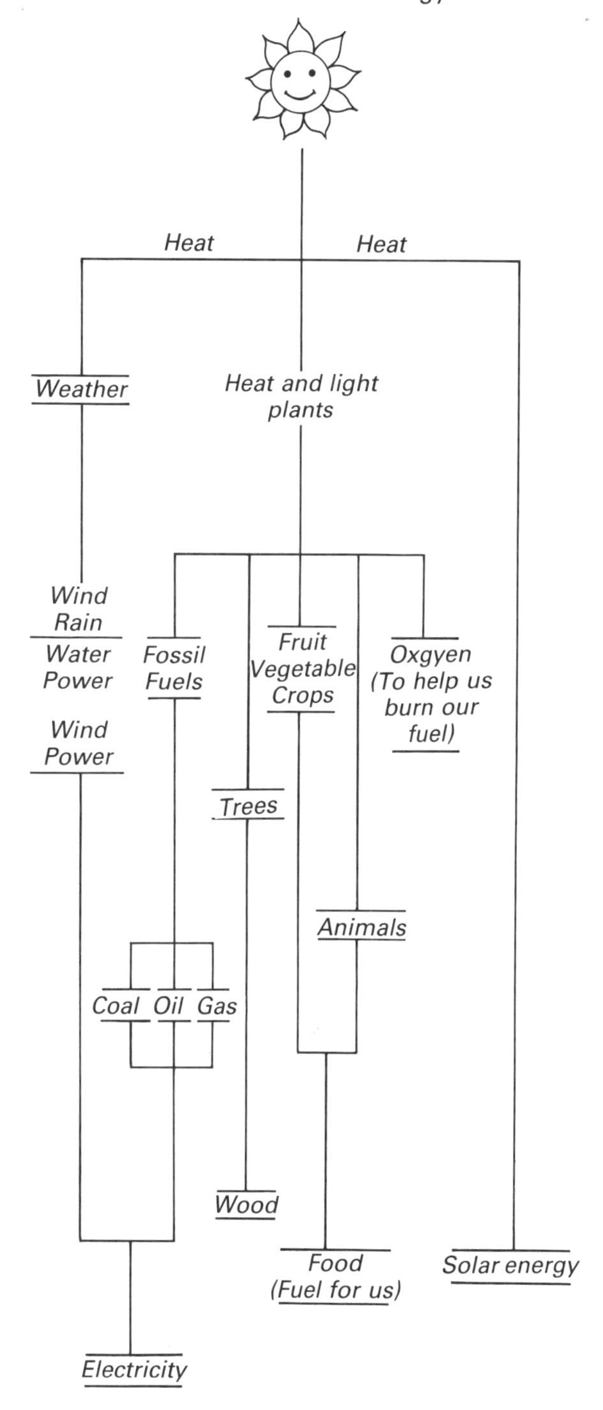

Investigating energy (2)

What happens when you cover a growing leaf with a match box for 2 or 3 days? The green colouring in a leaf shows that it is using sunlight to help convert (change) its food into stems and leaves. Why does the leaf go yellow?

We cannot store the warmth of the sun in a box but by burning wood we could release the energy which we feel as warmth.

Keeping some of the Sun's energy for use in the winter is a good idea.

Some other ways of storing energy are: petrol in a car's tank; electricity in a battery or water held in a reservoir for driving a turbine to make electricity.

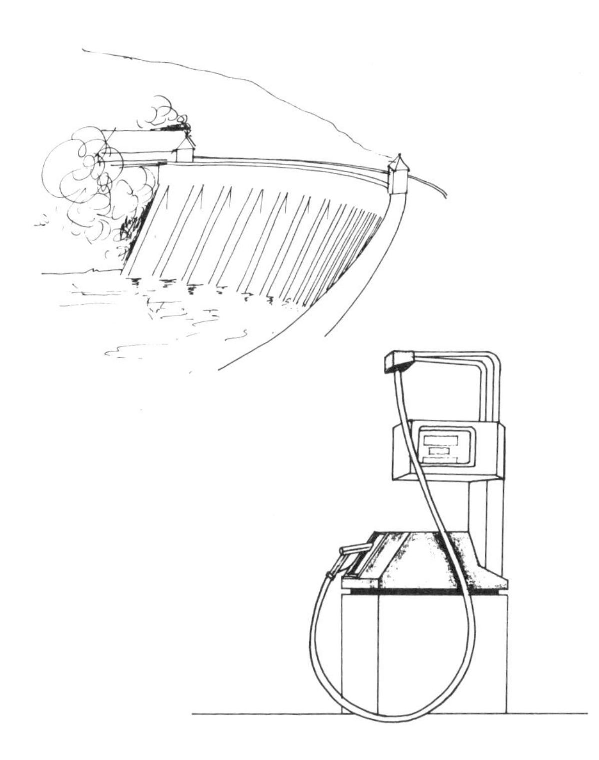

People have always tried to use the energy freely available from the Sun (solar energy), wind, water and tides. A solar panel traps the heat from the Sun and uses it to warm water. (See the illustration.) This warm water can help reduce the fuel needed to heat a house.

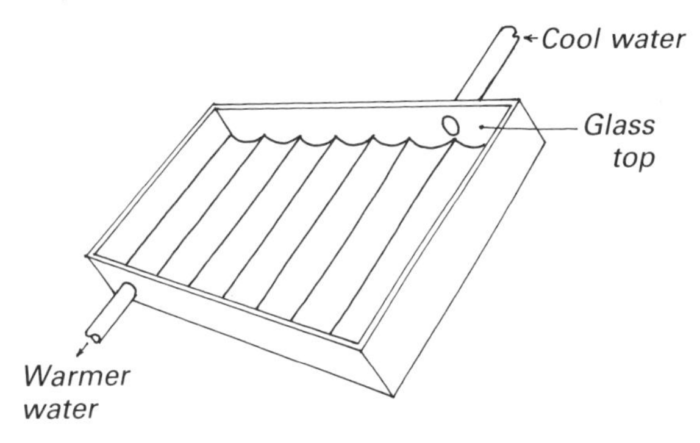

Which natural force has caused the change in the harbour? How do you think it would be possible to use the energy from this source?

When energy is stored and ready to do work it is called potential energy. When potential energy does work and causes movement it becomes kinetic energy. Heat, light and sound are all evidence that potential energy has been converted (changed) into another form of energy. Draw a diagram to show this conversion in each example below. An example has been completed for you.

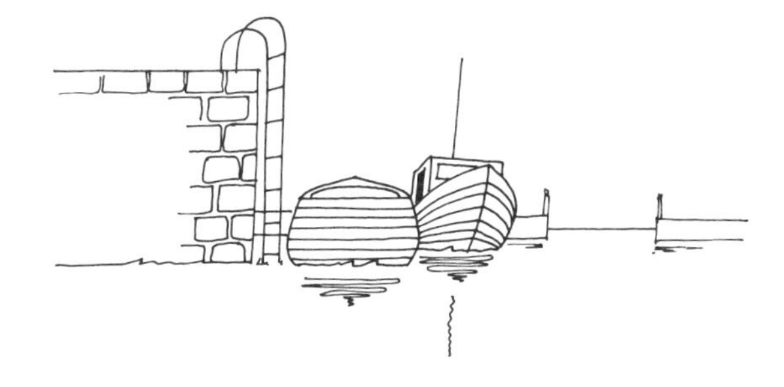

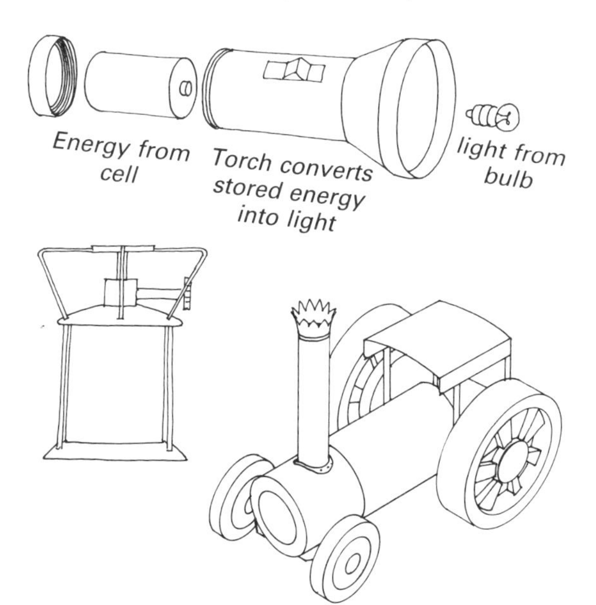

FRICTION

When things rub together energy is 'wasted' by friction. We cannot see friction but we can feel the warmth it makes when we rub our hands together. Some things get very hot when they rub together!
What other things get hot as a result of friction?

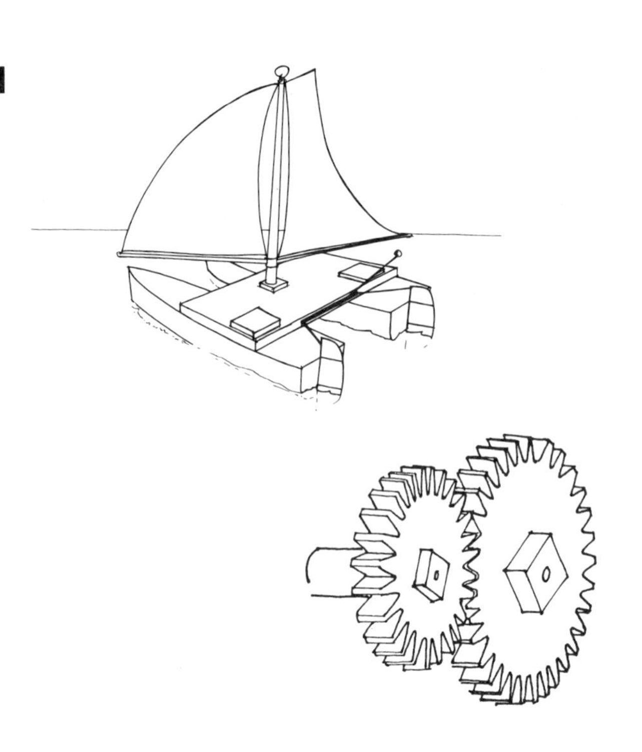

A windmill made it much easier to grind corn into flour. A lot of energy is used up by friction in old-fashioned windmills which is why the sails had to be so large.

Wind turbine generators in Orkney

Which form of transport uses sails powered by the wind?

Modern use of wind energy also makes full use of developments in bearings, lubricants and propeller designs to help generate electricity.

Investigating energy (3)

Energy from the wind varies with how hard the wind is blowing. Make a wind speed indicator to help you compare the wind strength on different days. The illustration shows one way of doing this.

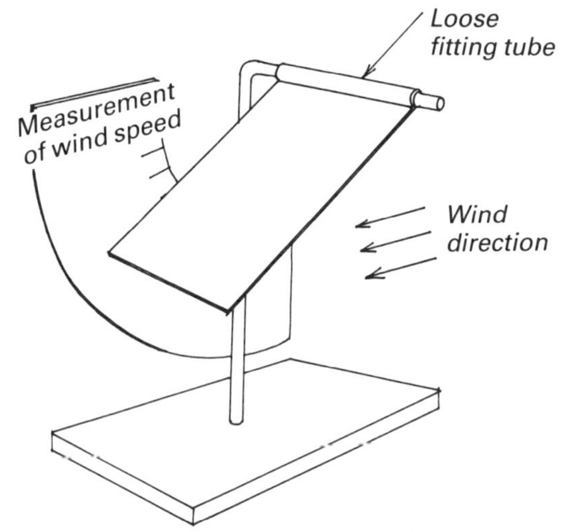

Listen to the coastal reports or watch the weather forecast and mark the scale with wind speeds. Can you devise another method for measuring the strength of the wind? What kind of problems are you likely to have when measuring the strength of the wind?

Energy from the wind can often be a disadvantage. A motorcycle fairing helps to reduce wind and air resistance.

Why is it unusual to fit a fairing to a pedal cycle?

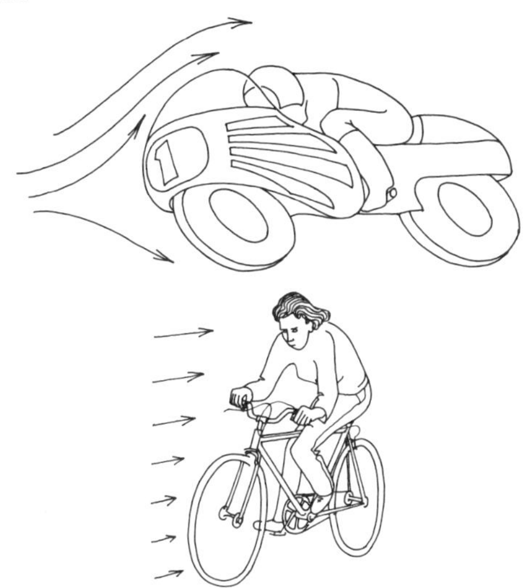

Until the twentieth century water wheels were an alternative to the windmill. They used the energy from a stream or river to grind corn or work machines.

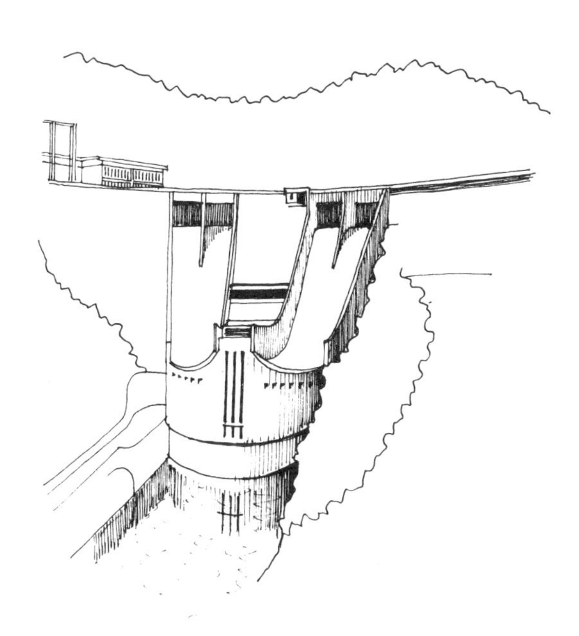

What can be done to a river so that it acts as a store of energy? What advantage does this give the watermill over the windmill?

The Sun is vitally important for plants and animals and it is from them that we obtain the fossil fuels—oil, coal and natural gas. Oil and gas formed when the remains of plants and animals which collected on the beds of oceans millions of years ago became covered and compressed by rock.

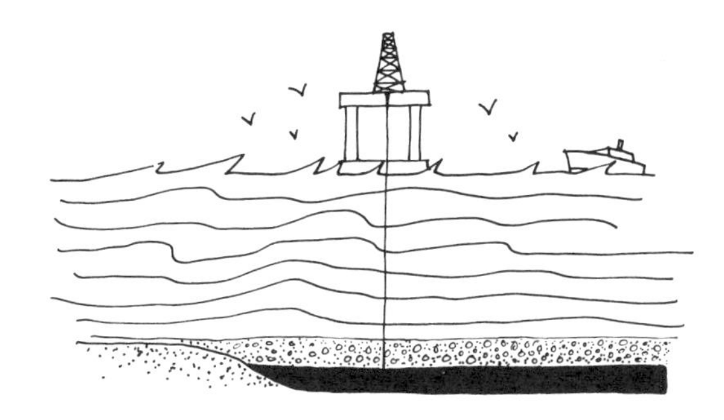

Coal is the remains of forests which became similarly covered and compressed by rock.

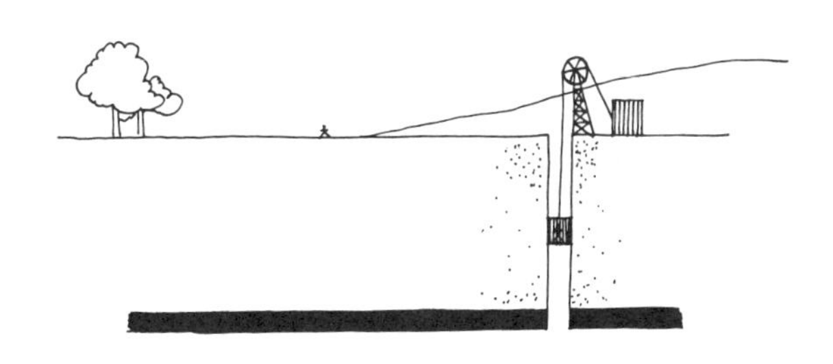

There is a limit to how long we can remove these fuels from the ground as they cannot be replaced. Nowadays there are fewer homes heated by open coal fires. However, coal is still widely used to generate electricity in power stations and for some central heating systems.

What three reasons can you think of for this change?

We are always searching for cheap (and hopefully everlasting) sources of energy. Nuclear power is a means of obtaining heat from the reaction caused by altering the natural balance within the atoms of certain substances. Whilst the heat from nuclear reactions is very useful for generating electricity there are possibly dangerous side effects which have to be carefully controlled.

Investigating energy (4)

A flow chart can show how energy is converted from one form to another. For example when someone is pushed on a swing the following are linked by energy changes. The Sun—plants and animals—food—children's muscles—swing raised (potential energy)—swing moving (kinetic energy).

Draw a flow chart to connect the Sun with the following:

(i) bowling a ball to a batsman
(ii) boiling an egg
(iii) using a washing machine
(iv) delivering the milk
(v) deep freezing fish
(vi) making paper

Some of these will have lots of branches which lead to the sun.

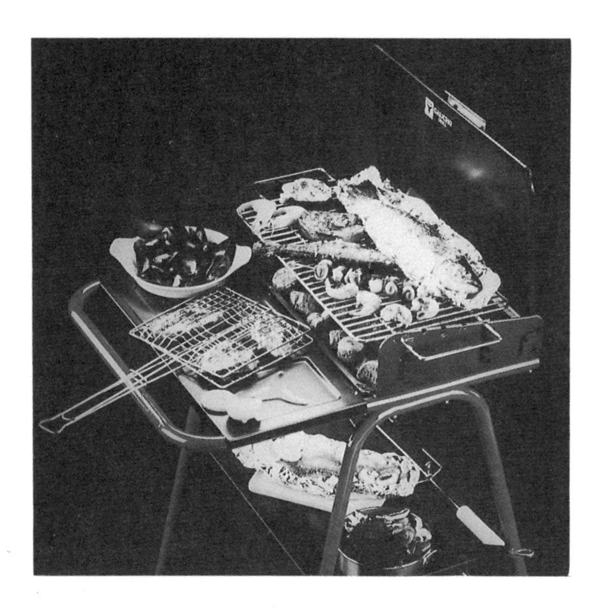

A barbeque uses the sun's energy stored in coke or wood to cook food

Investigating energy (5)

Think of ways in which you can easily convert:

(i) kinetic energy into heat energy
(ii) light energy into heat energy
(iii) potential energy into kinetic energy
(iv) heat energy into kinetic energy

ENERGY CONVERTERS

We have seen that although most energy comes from the sun it is available from a variety sources. One important source is gravity—the force which keeps our feet on the ground. Something which is raised above the ground can provide energy as it falls.

Energy project (1)

Design and make a timer which obtains its energy from a mass of 500 g raised 1 m above the ground. The timer must run for at least 30 seconds and if possible indicate periods of 5 seconds. The mass has potential energy before it is released and kinetic energy as it descends (See page 47) If the mass is suspended on a cord which is wrapped around a spindle the time taken for the mass to fall to the ground is very short—about 4 seconds. (The thickness of the cord is important and fishing line is recommended for its known strength and constant thickness). To make the weight fall more slowly you could try wasting some of the energy by making a fan brake similar to the one illustrated.

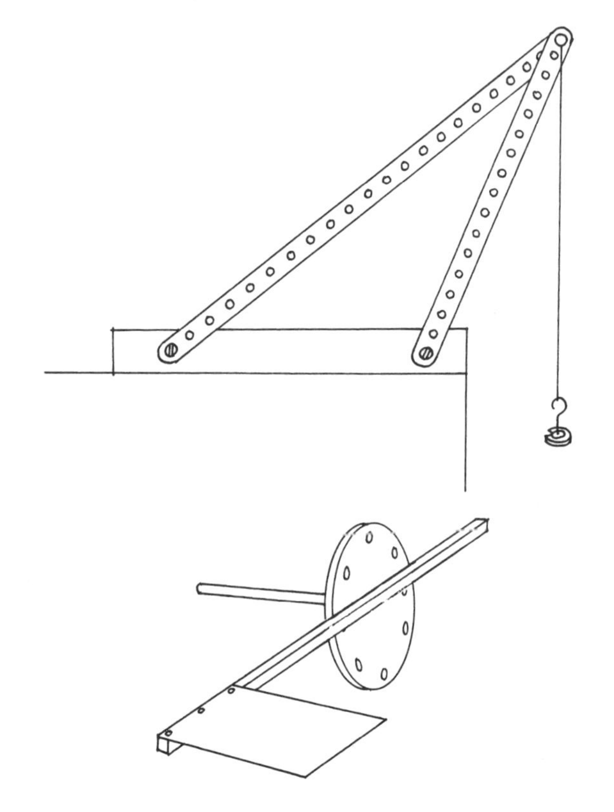

Trim strips off the cardboard fan until the spindle starts to turn. The timer can be made using a construction kit but cardboard pulleys with rubber band belts can also be used. Aim to make the last pulley in the system turn only once in the 30 seconds. You will need to refer to page 71 to find out how to work the size of the pulleys. The 'Service Section' of this book (page 86) describes how to make carboard pulleys.

Do not make any pulleys larger than 50 mm in diameter as it may be difficult to obtain very long rubber band belts. The spindles can be supported by bearings made from strips of tin plate or meccano screwed to a softwood base.

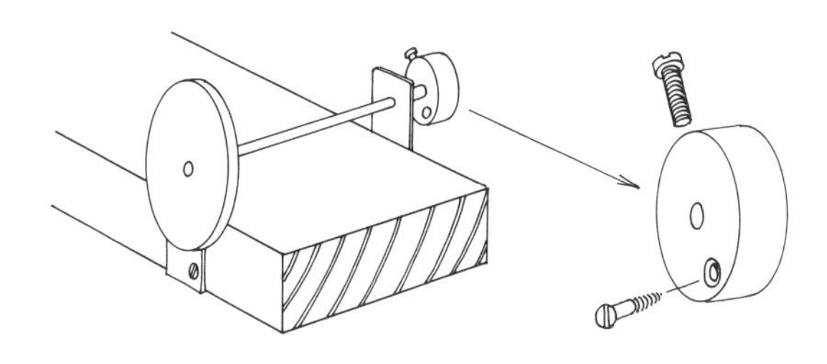

If you have a lot of gears or pulleys there may be enough friction at the bearings to be able to do without the fan brake. To measure seconds without using a watch count 'one banana, two banana, three banana' for the required number of seconds.

Elastic bands

An elastic band can act like a store of energy. Once it has been stretched (and so given energy) all it wants to do is get rid of it as quickly as possible. If we can control this release of energy then the band can do work for us. Do not stretch elastic bands and 'fire' them—this can be dangerous to other people! Compare the results obtained when different types of band are tested on the rig illustrated.

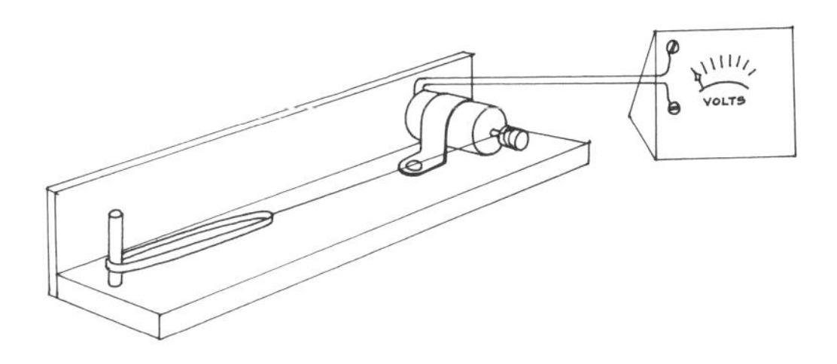

A multimeter can be used but it must be sensitive enough to give readings below 2 volts. Stretch each band to twice its normal length by winding the thread around the spindle of the motor.
Different types of band can be made by cutting slices from old inner tubes as shown.

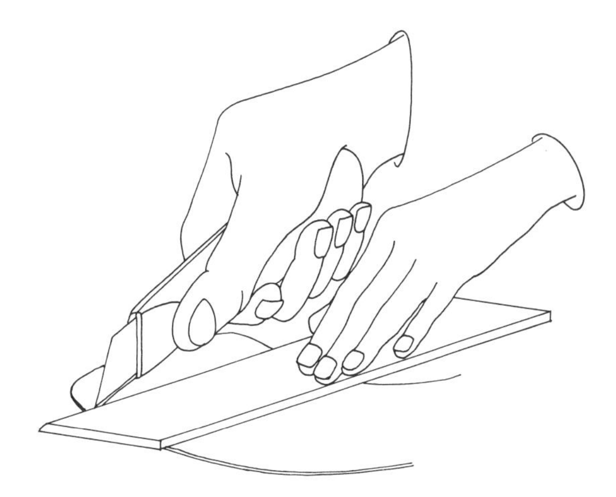

The other way of giving the band energy for it to store is by twisting it. Compare different bands on a test rig like the one shown.

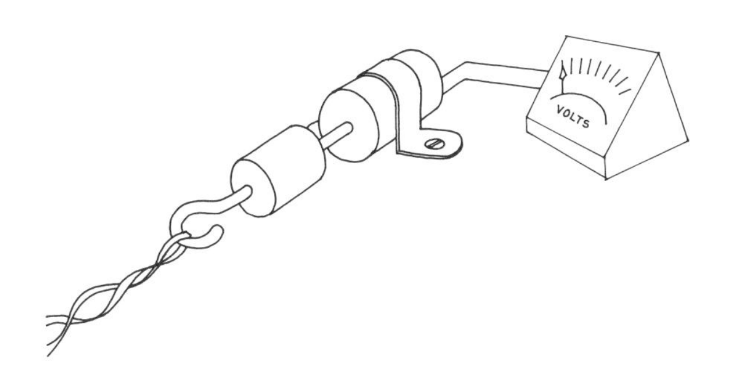

What happens to the readings obtained when you put a spot of oil or washing-up liquid on the twisted band? Read the section on friction to help you explain any differences that you might get in the results.

Energy project (2)

Design and make a vehicle to complete a 5 metre course as quickly as possible using the energy from an elastic band. If you are racing against other vehicles they must all use the same size of band or weight of rubber.

Use a construction kit to build a chassis (framework).

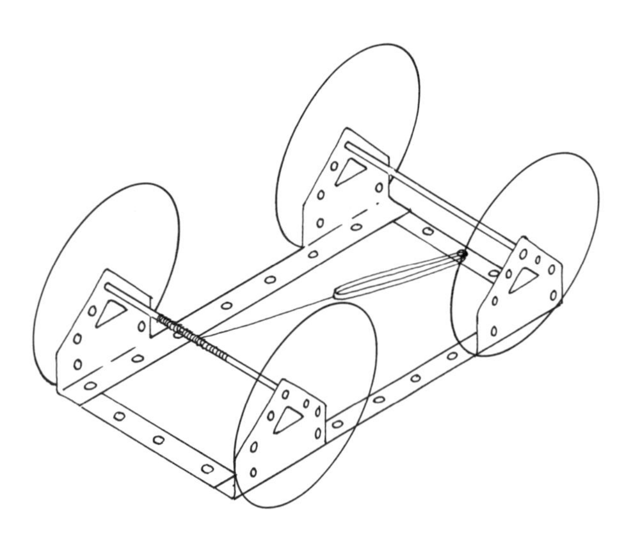

Rubber bands can be used to improve the grip of the wheels as they may tend to spin when the vehicle is released. Read the section on 'Mechanisms and control systems' to find how gears and pulleys may help to reduce the waste of energy when the wheels spin. The size of the wheels is quite important. Why do 'dragsters' (page 58) have large back wheels?

Clockwork motors

This kind of motor was invented as a simple and reliable machine to measure time. Energy is put in by a key turned by hand (the input) the key winds up a coiled spring which can be made to unwind slowly (as in a clock) or quickly (as in a toy train). The motion produced by the motor is called the output. The output of a clock is the movement of its hands.

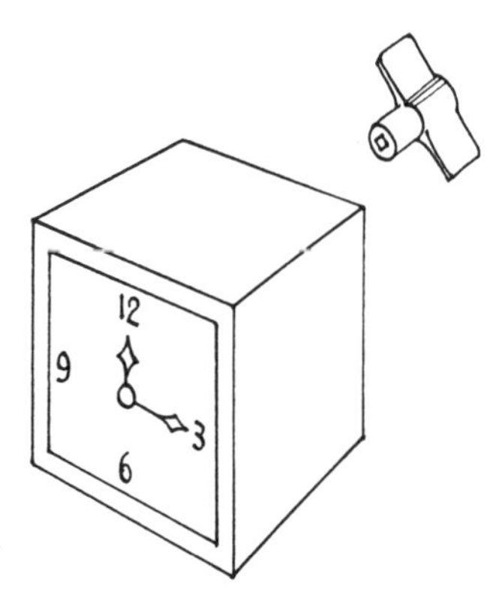

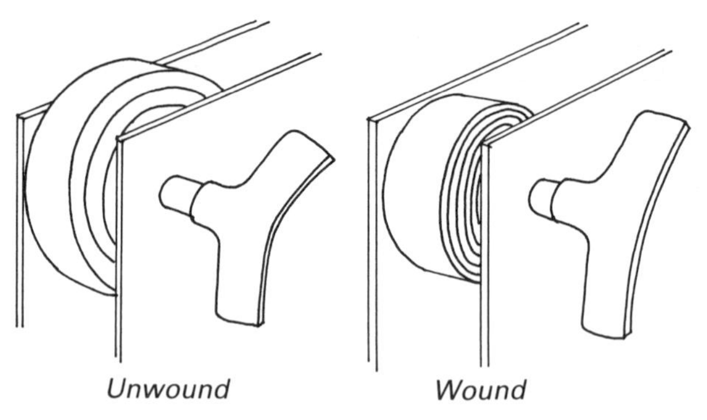

It is never possible for the input to be the same as the output as some energy is always wasted. When any two things rub together there is always friction which wastes energy.

Clockwork motors have lots of gears and spindles which produce friction. The section 'Mechanisms and control systems' in this book will help you to understand what happens inside the motor. The clock has to show the time accurately even when it is partly wound down. The output of the motor is measured in revolutions (turns) per minute. How quickly do records revolve on a record player?

Things which turn very slowly may have their speed measured in different units. The world turns once per day. A clock's minute hand turns once per hour which could be written as a 1/60 rpm and this has to be constant.

A stroboscope is an electrical machine which flashes a light accurately at a revolving object and can be used to indicate how quickly the object is spinning.

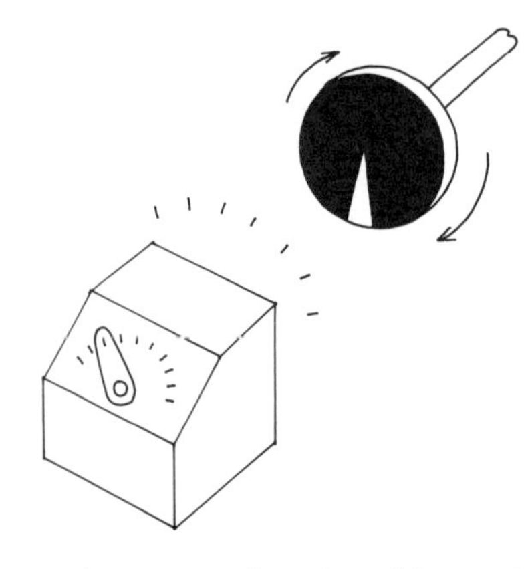

Using a stroboscope: when the white mark appears to stand still a reading can be taken

Even without a stroboscope you can accurately check the speed of a motor. Fasten a piece of 20 mm diameter dowel onto the end of the motor spindle and tape the end of a long piece of thread to it.

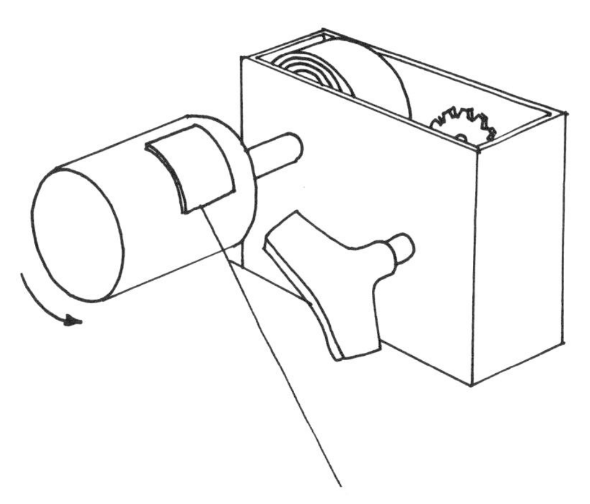

You will need a watch to measure accurately in seconds. Run the motor for 5 seconds and then count the number of turns of thread that have wound onto the dowel. If the motor runs for 5 seconds and winds 30 turns onto the dowel then it is possible to calculate the motor's speed as follows:

in 5 s the spindle turned 30 times
in 1 s the spindle turned $30 \div 5$ times (6 rev/s)
in 1 minute (60 s) the spindle turned
$60 \times 6 = 360$ rev/min

Compare the speed of the motor when it is fully wound, half run down and about to stop. The way in which the motor runs down could be important when doing a project like the one described below.

Energy project (3)

Design and make a model of a lift which is to be used in an explosives factory and must move dangerous chemicals as slowly as possible. The load in your model is to be a maximum mass of 50 g and it is to be lifted through 750 mm. Use a clockwork motor as a source of energy. You could use the fan brake from the timer project (page 50) or you could make a band brake, as shown in the diagrams to help slow down the output motion of the motor. The output speed may change as the motor runs down or as it is loaded. If you do make a band brake the spindle must run easily in the bearings and the drum must be as smooth as possible.

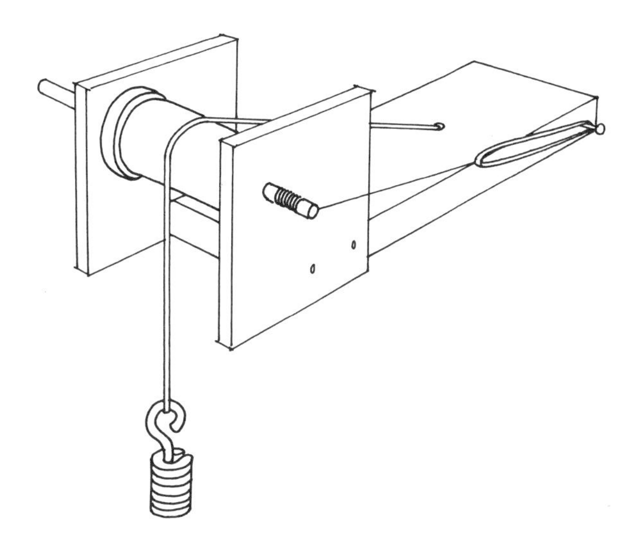

Electric motor

The electricity needed to power an electric motor can come from a battery or a transformer which converts the mains voltage to a lower level more suitable for use in schools. If you are using cells or batteries you should test them using a multimeter. Set the meter to the 'd.c. volts 1–10 range' and connect the leads to the terminals of the battery as illustrated.

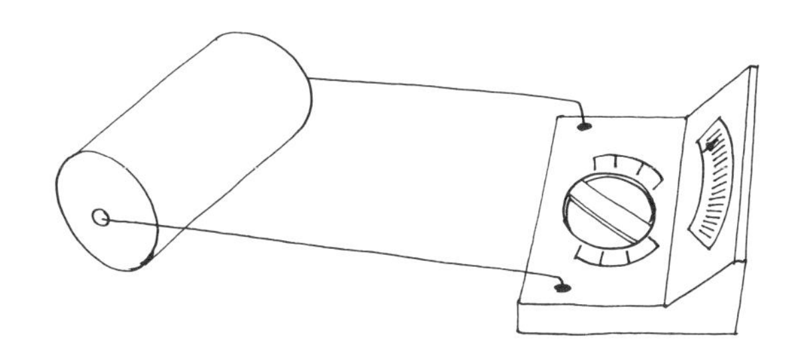

The meter will have ranges marked d.c. and a.c.; d.c. means direct current which is the type obtained from a battery. Electrical appliances like vacuum cleaners work on a.c. What does a.c. mean?

We have already shown (page 51) that if you rotate (turn) the spindle of an electric motor then you produce electricity (like a bicycle dynamo). An electric motor more usually has an electrical input and a mechanical output. The two main parts of a motor are a coil of wire and a magnet. When electrical current passes through the coil a magnetic field is produced. The magnetic field reacts with the motor's magnet to cause the spindle to turn.

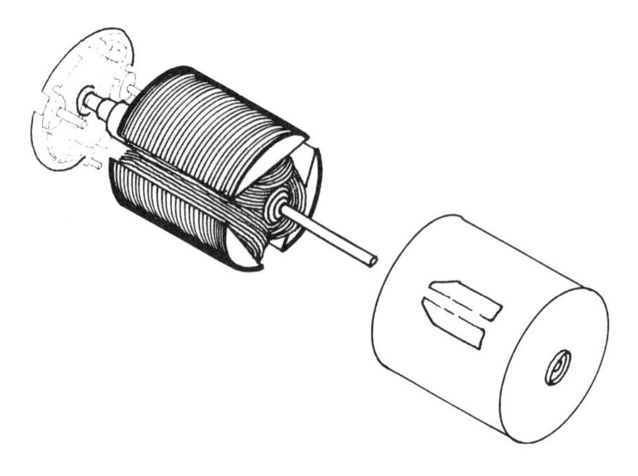

Because the electrical current passing through a coil also causes heat, deliberately stopping the spindle (or stalling it) whilst the motor is connected to a power supply, will damage the motor. For this reason never use a fan brake to reduce the output of an electric motor.

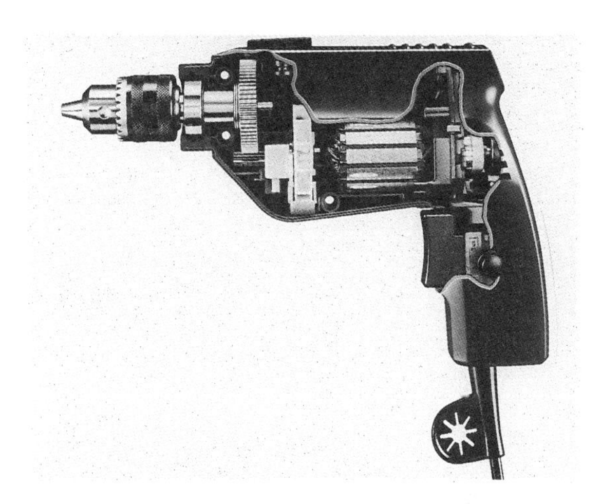

The fan inside this drill does use some of the drill's output but it is there for a special purpose.
Can you explain why it is there?

Energy project (4)

Design and make a hill-climb vehicle using a 9V d.c. motor which can carry its own battery or be supplied with power through trailing wires. The vehicle is to climb the steepest possible gradient (slope). Speed is not important. You will need a ramp on which to test your vehicle. The ramp needs to be adjustable in order to give different gradients. Because motors run most efficiently at full speed it will be necessary to use gears or pulleys so that the high rotary speed of the motor spindle can be changed to the low speed that the wheels should have. See the Mechanisms section of this book for information on gear ratios. Use a construction kit or make cardboard pulleys (see Service section page 86) to obtain the required speed reduction. Do not make any pulleys larger than 50 mm diameter as you may have difficulty in obtaining elastic band belts large enough.

CONSERVING ENERGY

Energy is too valuable to be thrown away. There was a time when oil, coal, wood and other natural sources of energy were very cheap and wasting some did not seem to matter. We are now more aware of how wasteful we used to be in the past and how we should conserve (keep) more energy for the future.

A 'Puffing Billy'

Scientists sometimes talk about the principle of Conserving Energy which is a way of showing how well a machine works for us. Make sure that you understand how the word 'conserving' is being used when you read it. This part of the book is about using energy properly. If the heat from an electric heater or a gas cooker is wasted, then the person who pays the bills is wasting money as well as energy.

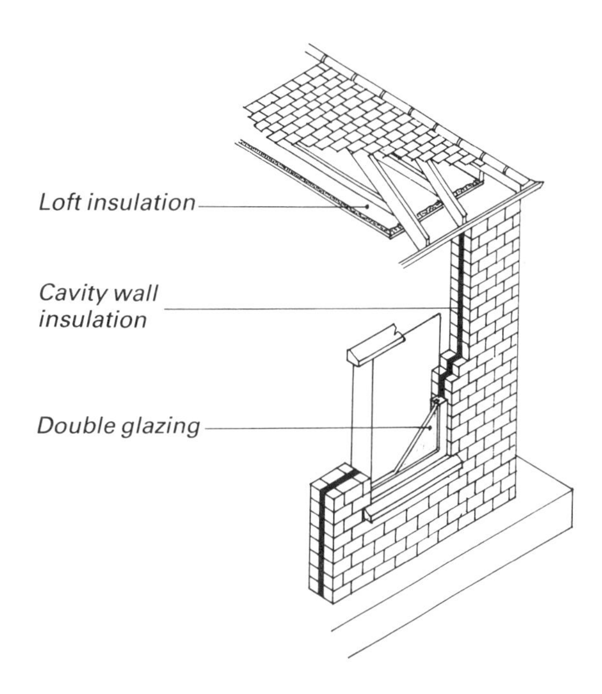

Write a description of the kind of clothing which relies on the use of air as an insulator.

Modern houses are often built so that they cost less to keep warm—they are more Energy Efficient. Owners of older houses may have to put in double-glazed windows, loft insulation and cavity wall insulation. Each of these is a means of trapping air which is a good thermal insulator.

Does your home have insulation to save heat?
Why are waterpipes often insulated?

Draw a diagram of a hot water storage tank to show how heat can be prevented from escaping from its sides.

Industry and transport use a lot of energy and much time preventing the waste of this energy.

The underground system in Glasgow is streamlined and highly automated to save energy

Conserving energy saves money and makes precious energy sources such as oil last longer. By careful planning it is sometimes possible to take hot water that would have been wasted from a factory and transport it to somewhere that can make good use of it. Some housing estates use hot water like this to run their central heating systems. Fish farms and people needing heat for greenhouses can also benefit.

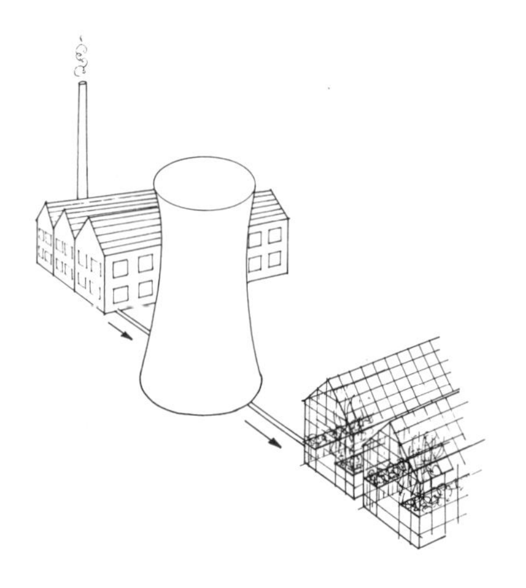

Read page 6 again. What types of energy are being used? Is it possible to conserve energy which used to transport people around?

Food gives us our energy but we waste an awful lot. Look at the way people leave food in a cafe, restaurant or when eating school dinners. Where does the waste food from places like hospitals go to? It is not always put into rubbish bins.

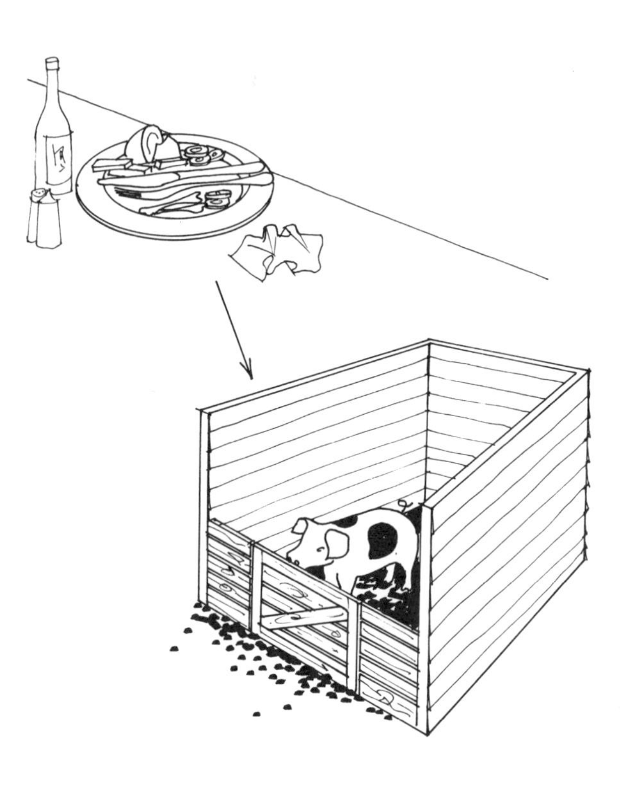

The next time you peel a banana or an orange think how much it cost to transport just the peel from the place where it grew. The best place for natural waste such as potato peelings is a compost heap which provides a source of energy for new plants.

When something is used again it is called recycling. As energy costs increase it becomes more worthwhile to re-use some materials which cost a lot to make.

What happens inside a compost bin or heap to break down vegetable and animal waste?

There are a number of places where energy can be wasted in the home. Anywhere where heat is produced there is a possibility that it can be wasted.

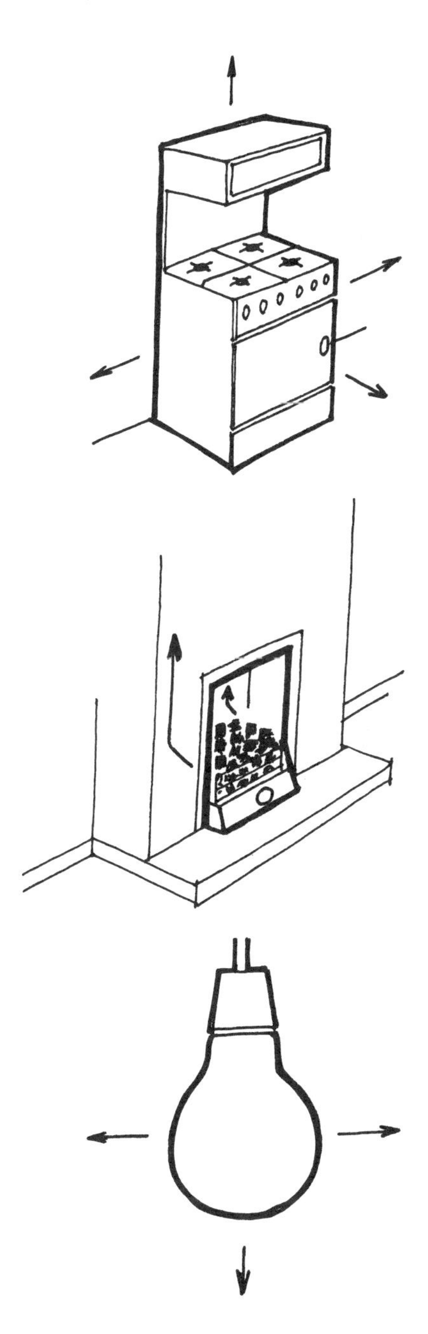

Glass needs a lot of energy to make and old bottles are often collected at bottle banks for recycling. Paper can be recycled to make packaging such as egg boxes. Scrap or waste merchants perform the vital task of collecting material so that they can be re-used.

What else does a cooker heat besides food? An open fire looks nice but where does a lot of the heat go? A light bulb produces heat as well as light. Do we need that heat? After all, we pay for the electricity that makes it.

Control systems

Energy has to be controlled for us to be able to make use of it. Kinetic energy is often controlled by mechanisms.

MECHANISMS

The photograph (right) shows a bicycle—a machine with which you will be familiar. A machine is any thing which reduces the amount of force that is needed to do a particular job. A screwdriver is a machine because by using it you need less force to turn a screw than you would if you were holding the screw head in your fingers.

Is a nutcracker a machine? We will find out the answer soon.

The bicycle may not seem very technological but it is when compared with early wheeled vehicles. The illustration is just for fun but it may make you think of how important the wheel has been throughout history.

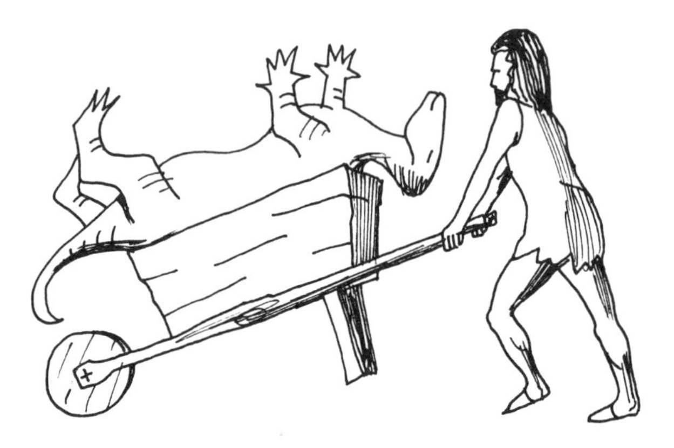

Make lists and sketches to show how and where wheels are used in different forms of
(a) transport
(b) industry and manufacture and
(c) where they are used in the home

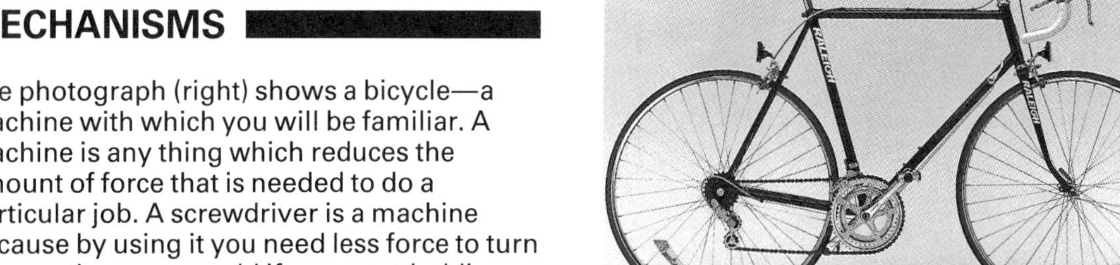

At one time machines had to be made with crude tools. The first of these machines are sometimes referred to as the five ancient machines which are the wedge, inclined plane, screw, lever and windlass. The wedge is used for forcing things apart and squeezing things together. The illustration shows a wedge being used to split logs.
Which would be easiest to force into the wood, a thin tapering wedge or a short fat wedge?

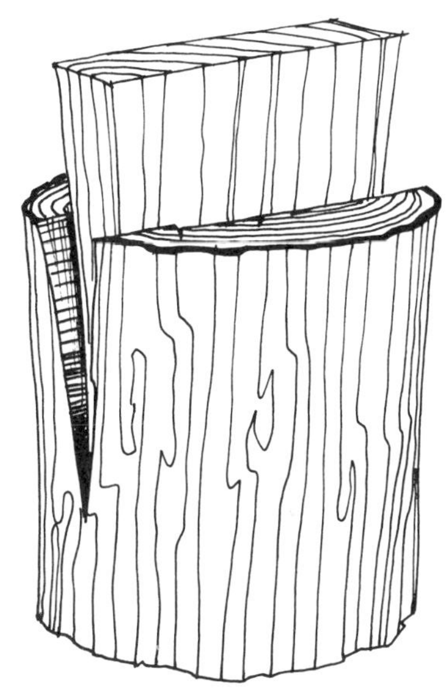

There are many tools used for cutting materials that use the wedge to make them more effective. An axe is more convenient for splitting logs than a wooden wedge.
Draw a diagram showing each of these tools being used and indicate where the wedge is for each one.

A chisel (take care it has a sharp edge)
A plane
A twist bit for a drill. You may find this difficult to draw but try to find the wedge.

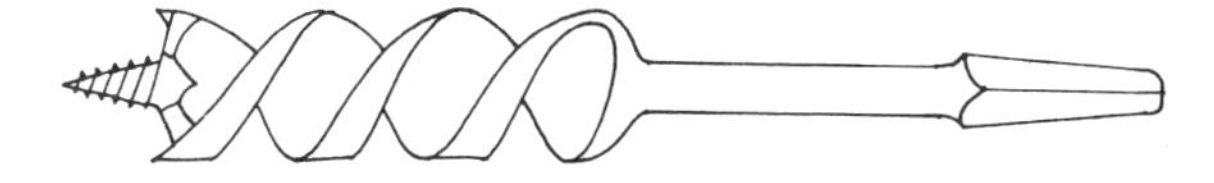

Investigating wedges

How could you compare the two types of wedge by using masses, blu-tak or plasticine, rule and timer?

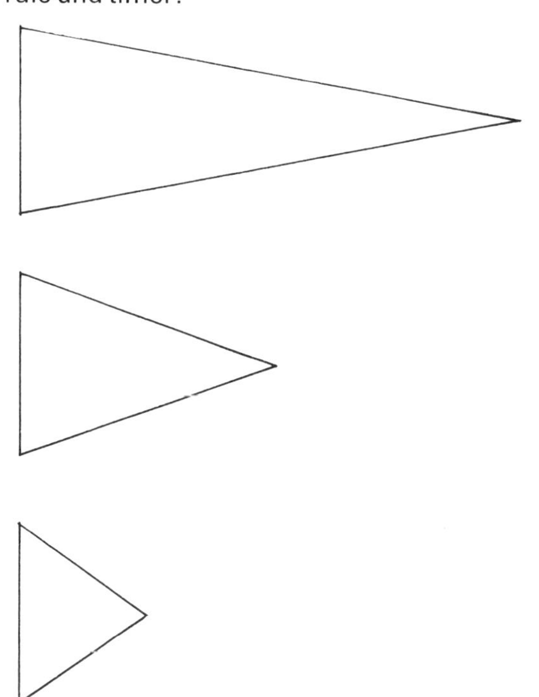

A wedge can also be used to hold things together.

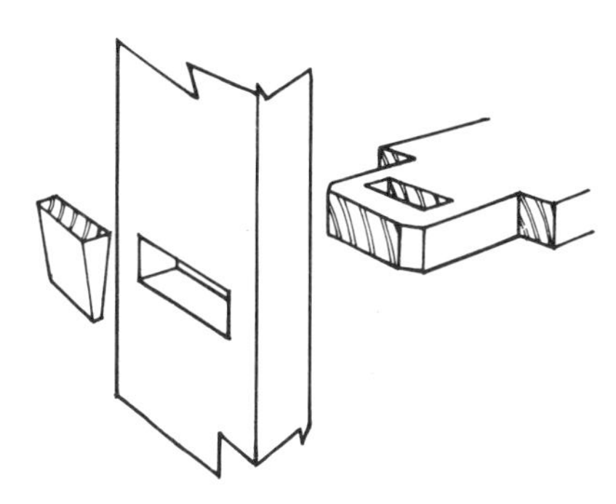

Simple tusk tenon . . .

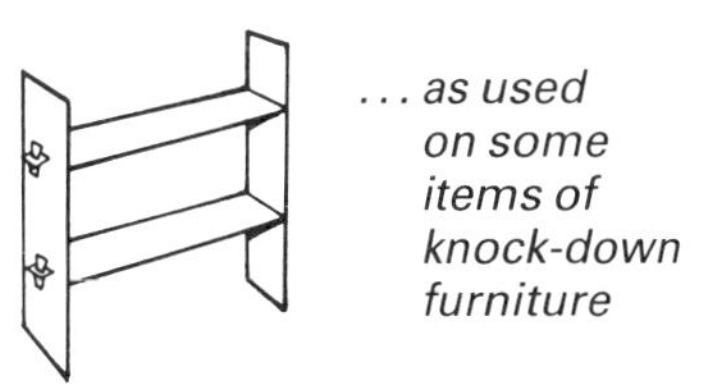

. . . as used on some items of knock-down furniture

When a heavy object such as a wheelbarrow has to be pushed up a step the wedge can be used to help.

Used in this way, it is called a ramp or an inclined plane. 'Inclined' means sloping and 'plane' means a flat surface.

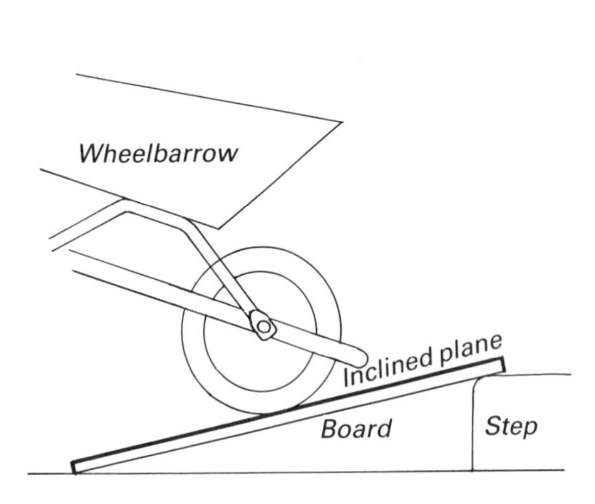

How does a wedge or inclined plane work? To find the answer count the number of steps in a staircase at school or at home. Imagine how much more difficult it would be to get from one floor to the next if there were half as many steps that were all twice as tall.

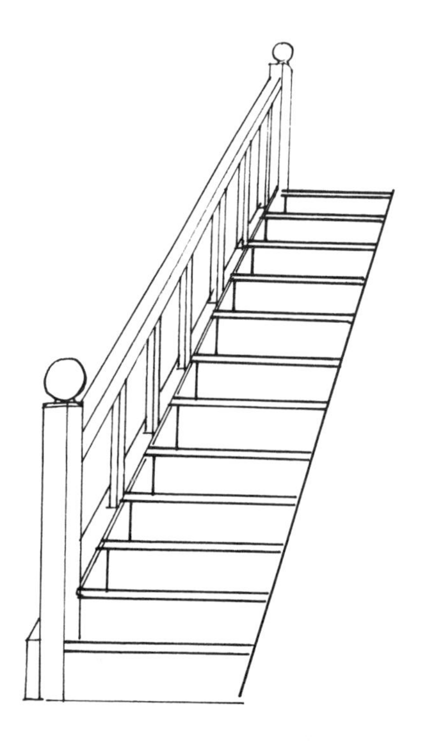

How difficult would it be to climb the same distance if there was only one step halfway between one floor and the next?

If a number of small steps is easier to climb why are there not more steps in our staircases?
The answer to the last question is that if we had more steps in the staircase and each step is big enough for our feet then the space taken up by the staircase would be very much greater. There is a limit to the space that is taken up by flights of steps.

One way around the problem is make the steps go around in a curve.

A spiral staircase

A spiral staircase allows people to climb the same distance but by using less floor space. When an inclined plane is wrapped around a cylinder (tube shape) the spiral line that is made is called a helix. The hand-rail of a spiral staircase is a helix.

Investigating the helix (1)

The illustration shows a flight of steps. By using a scale to reduce these large sizes we can draw a diagram of the steps on paper.

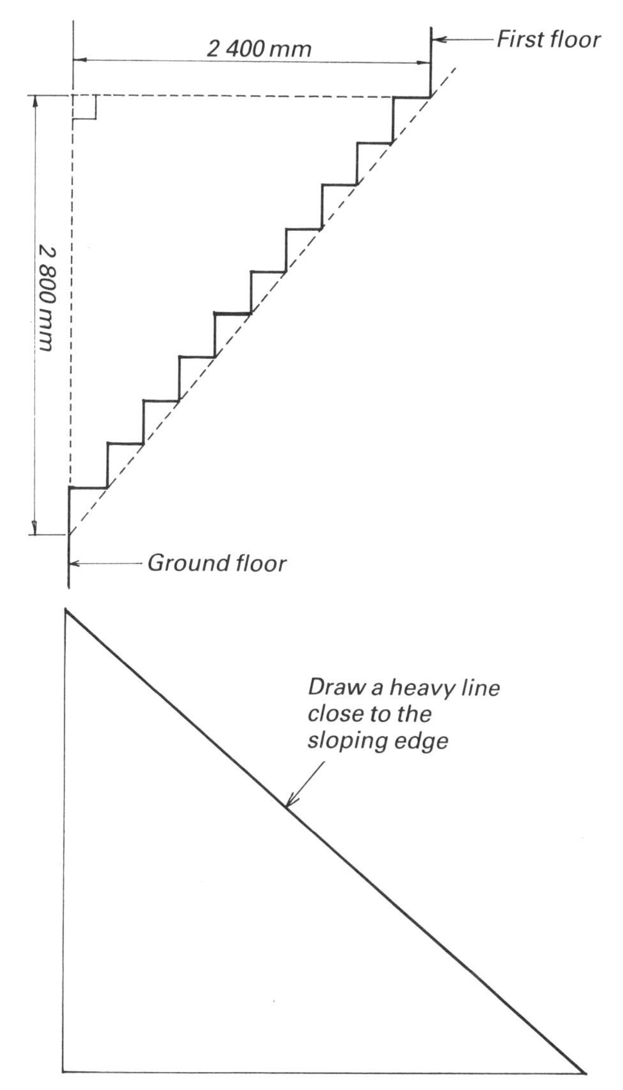

A convenient scale would be 1 mm representing 20 mm (this is written as 1 : 20 and read as 'one-to-twenty'). Draw the triangle shown in broken lines (do not draw the steps) and measure the angle. Most steps and stairs are no steeper than 42°.

Cut out the triangle and wrap it tightly around a pencil, keeping the base of the triangle level.

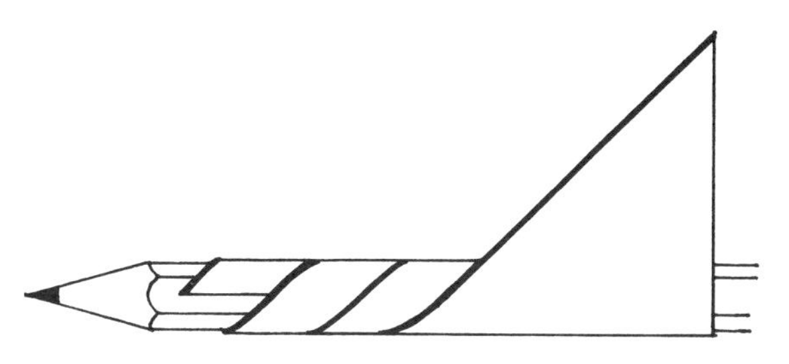

The heavy line will now be a helix. If this helix and pencil were the middle of a spiral staircase how many times would somebody go round in a circle in order to climb them? (Keep the paper tightly wrapped around the pencil.)

You should have counted the spaces between the lines. People do not like going round and round too much, it makes them dizzy. Allow the paper shape to unwind so that it can stand up on its own.

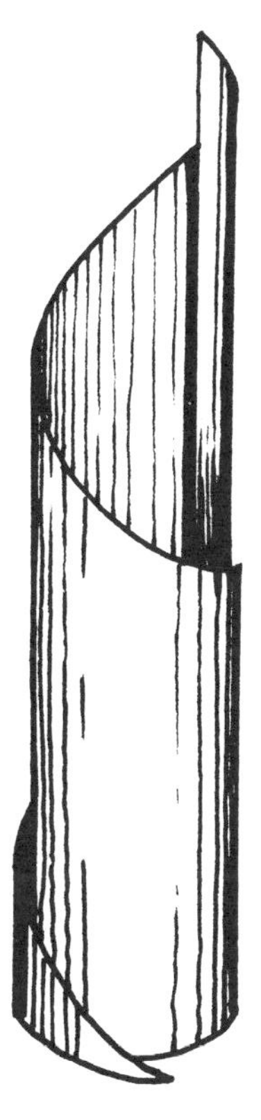

Now count the number of times that someone would have to go around while climbing this helix. Is this number more acceptable? If not how could you make it better?

The space between the parts of the helix that you can see at any time is called the pitch. When the helix is wrapped tightly around the pencil the pitch of the helix is about 23 mm.

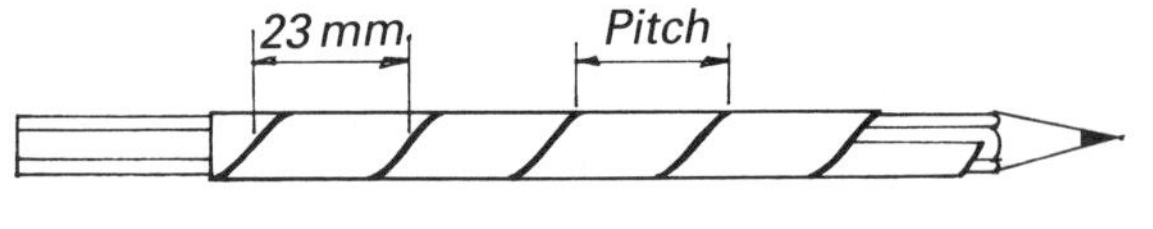

Tightly wrapped

What is the pitch when the paper is allowed to unwind and stand on its own?

THE SCREWTHREAD

When nuts and bolts are made they have an accurate helix shaped groove formed onto them. Nowadays the grooves are made so accurately that all nuts will fit all bolts of the same type. There was a time when a nut would only fit the bolt that it had been specially made for.

A handmade thread

Look carefully at a nut and bolt with coarse metric thread (M8 × 1.25).

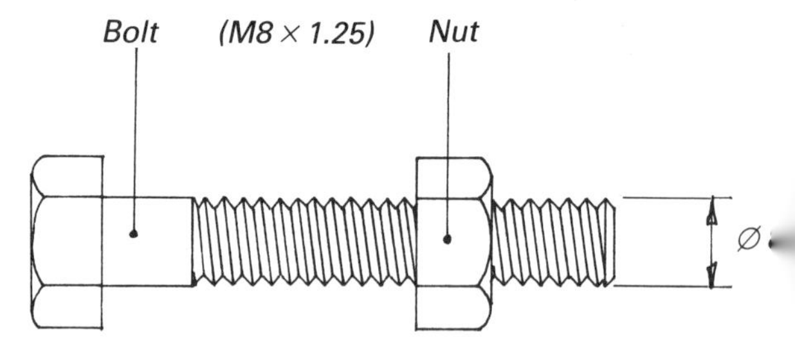

'M8' means a metric thread which is 8 mm in diameter. '1.25' is the pitch of the thread. Every time the nut is turned once on the thread it will move a distance of 1.25 mm along the bolt. This distance is very small. How could you check the pitch of a thread (approximately) by turning the nut 20 times and measuring with a rule how far it has travelled along the bolt?

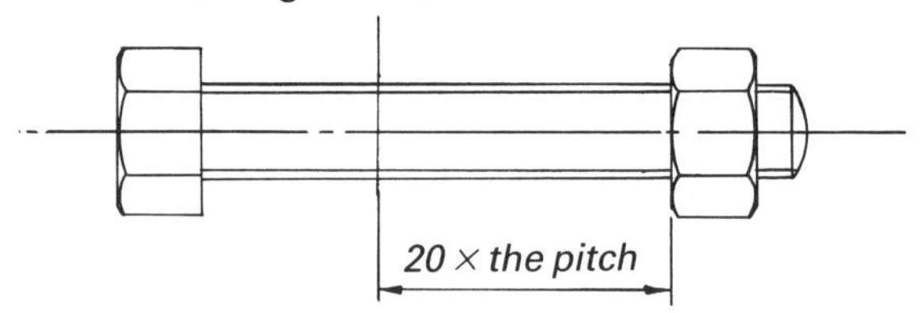

The pitch of the thread is small because the helix is tightly wound. When the nut is turned it can easily squeeze parts together because the helix was made from a long thin triangle. Look at page 59 which describes different wedges. Short fat wedges need more energy to do the work of a long thin wedge.
The photograph shows a jack which can lift heavy weights. Is the pitch of the thread large or small?

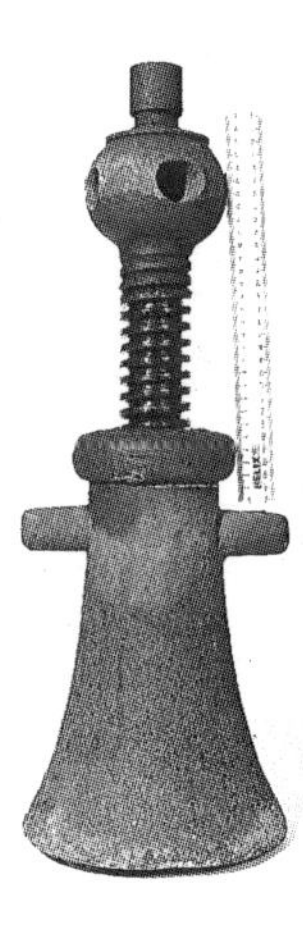

Investigating screwthreads

Find as many examples as possible of parts which are held together by threads. Some common things are coffee jar lids, squash bottle caps, toothpaste tube caps, shampoo bottle tops as well as many different sorts of bolts and screws.

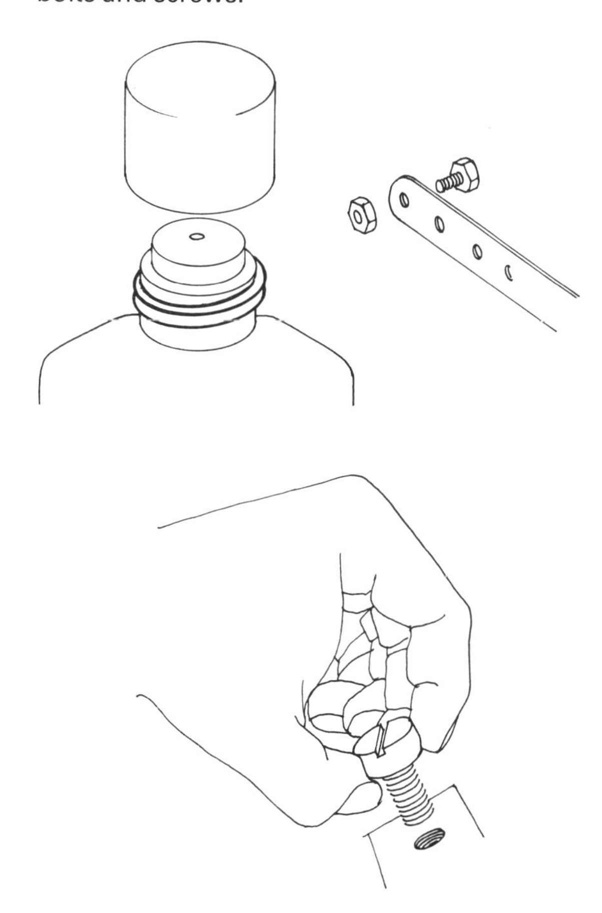

What is the difference between the bottle cap thread and the others shown here?

Use a rule to measure the pitch of the thread. Try to find things which should be using a screwdriver or spanner. Do not take things apart without permission. Are the pitches of these threads bigger or smaller than the caps that are meant to be taken off by hand?

Investigating the helix (2)

Write a description of the way in which the helix is used in the following buildings: lighthouses; castles; churches; and multi-storey carparks. Use diagrams to illustrate your description.

LEVERS

Earlier we said that one of the earliest examples of technology was when a branch was used to move a rock.
Does using a branch in this way make it a machine? If you do not know, read page 58 again. We do not often use branches to move things nowadays but we do use spoon handles to remove treacle tin lids, screwdrivers to open paint tins and planks to make see-saws. All of these are examples of levers. The easiest to understand is the see-saw.

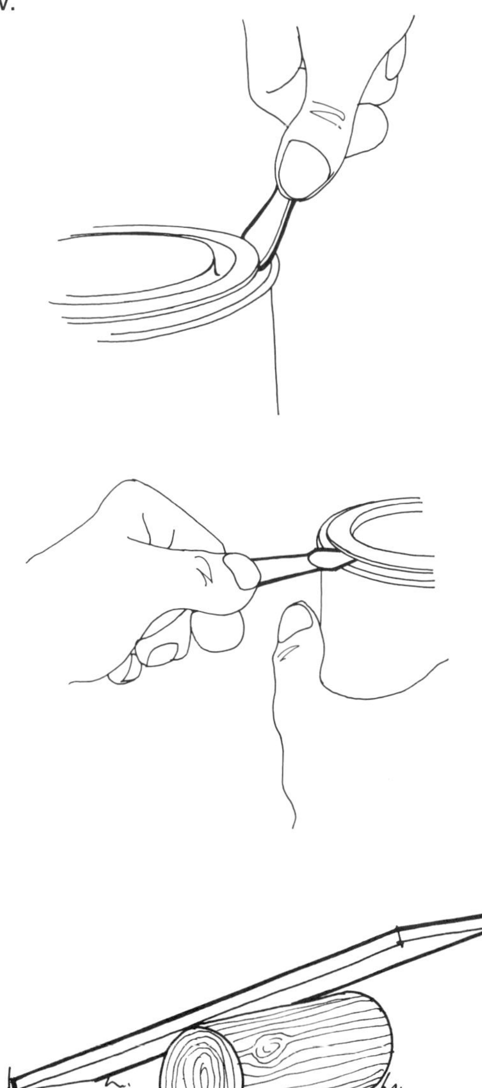

A simple diagram of the see-saw is shown below.

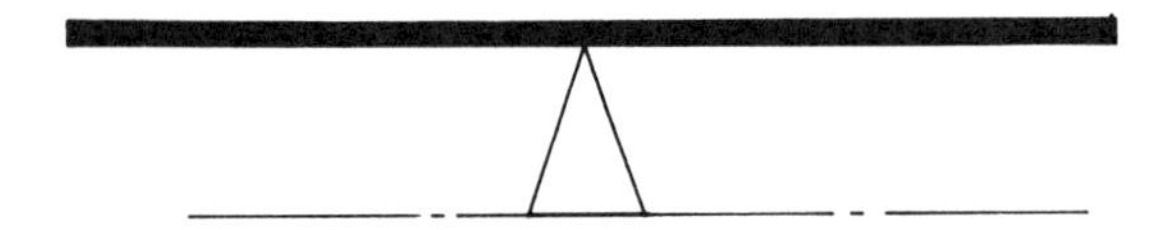

The see-saw balances at a point called the fulcrum. When two children of the same weight sit on the see-saw (one at each end) the see-saw is balanced. Another way of saying 'balanced' is 'in equilibrium'.

Investigating equilibrium

Build the framework shown in the diagram. The easiest way to do this is to use Meccano.

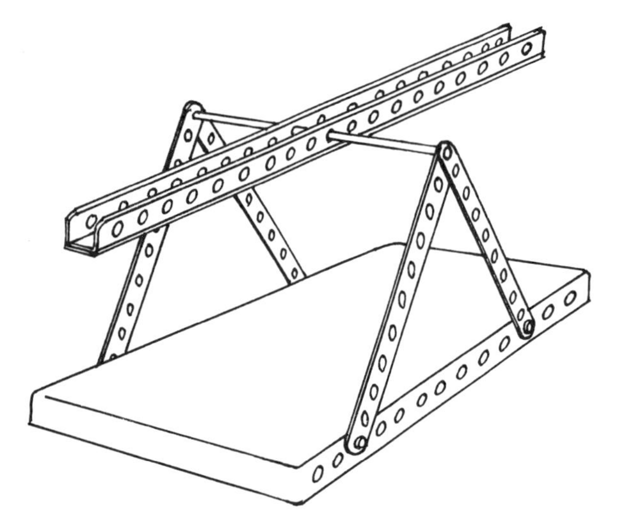

Use the longest strips or angles to make the lever.

Start with the axle passing through the centre of the lever and find out how well it balances. Hook a 100 g mass hanger into the holes at the end of the lever. You may find that it is difficult to get a perfect balance. Why? Move the fulcrum so that it is three holes to the left of the middle.

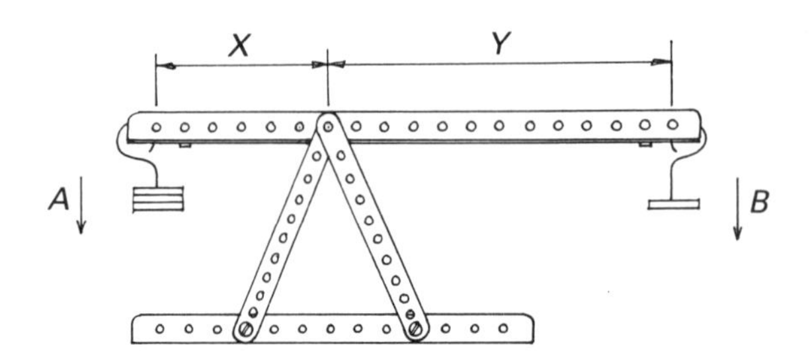

Make the lever balance by adding masses to one of the hangers. Measure the distance (in metres) from the left-hand mass to the fulcrum and write it down. Also write down the size of the left-hand mass (in newtons). Repeat this for the right-hand side. When you multiply the left-hand mass by the left-hand distance you will get an answer in newton/metres, e.g. $X \times A = XA$ newton/metres.

Multiply *Y* and *B*. You should find that the result will be the same as for the left-hand side. The investigation should show that the left-hand moment (moment = mass × distance to fulcrum) is equal to the right-hand moment when the lever is in equilibrium.

A steelyard

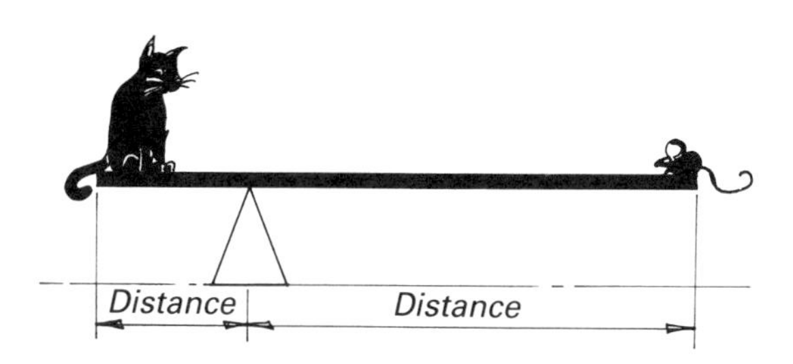

Left-hand mass (cat) × distance = right hand mass (mouse) × distance. All kinds of machines use different levers but they all have a fulcrum and moments.

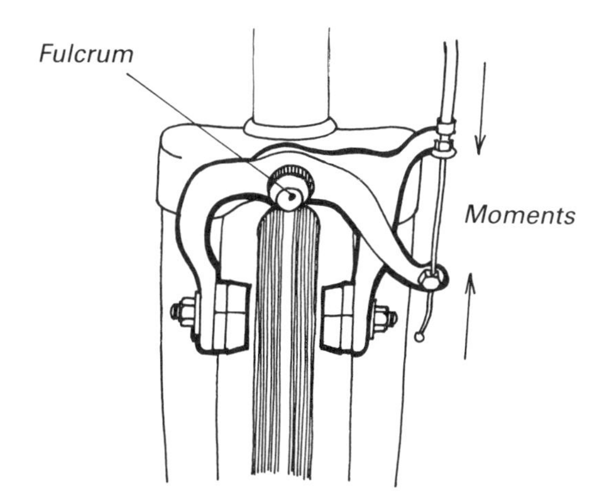

Bike brakes

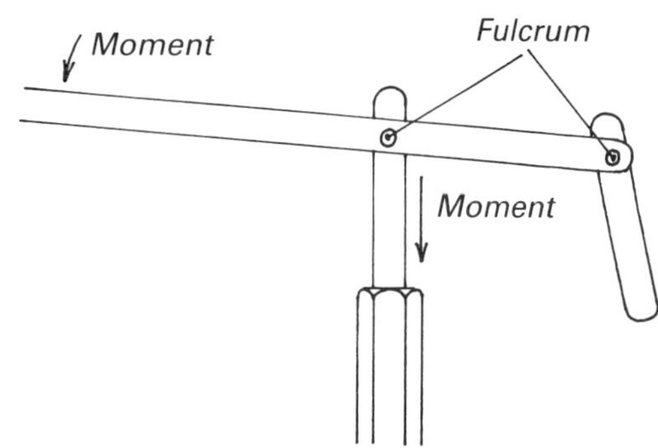

The lever system for a car jack

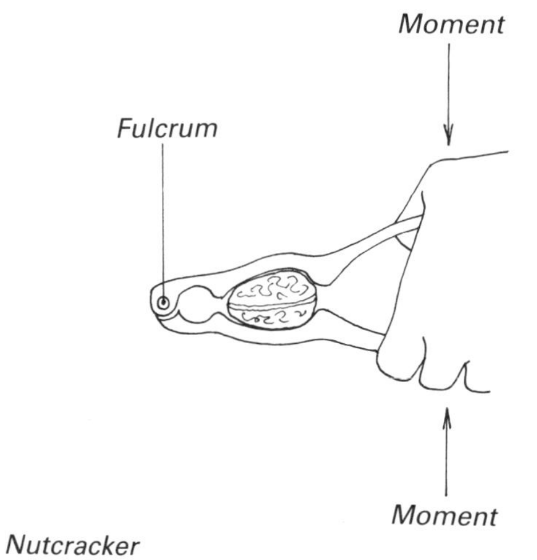

Nutcracker

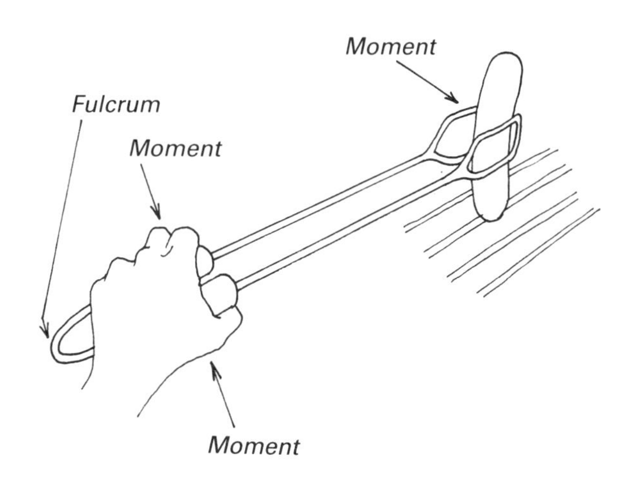

Barbecue tongs

Examples of different types of 'orders' of lever

Investigating moments

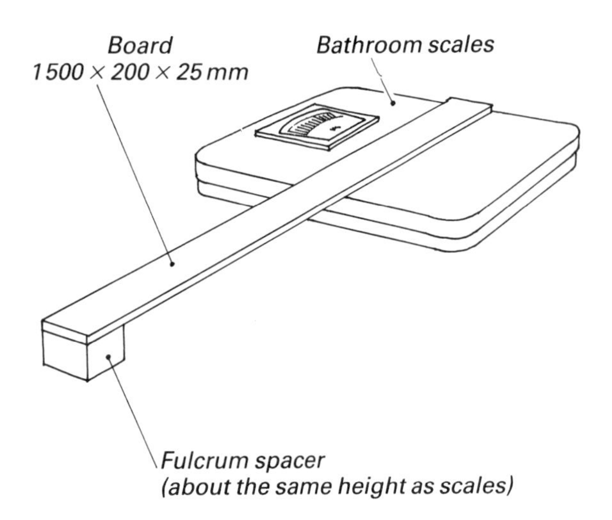

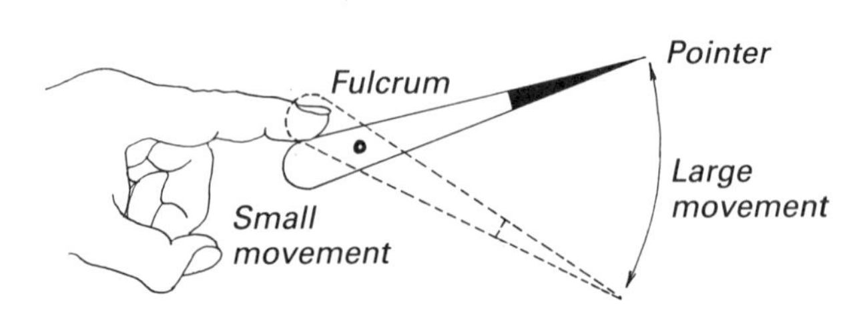

Stand on the board on the scales and weigh yourself. Slowly move towards the fulcrum and watch what happens to the reading. Find a position on the board which gives a reading of half of your weight. Find a position on the board which gives a reading of 10 kg on the scales. How much load are you putting on the other end of the board?

Moments project (1)

Design and make a device which will give readings of up to 5 kg in steps of 0.5 kg. You can use any position for the fulcrum for the lever(s). You could use a system similar to that described in the last investigation but much smaller. Use a pad of upholstery foam instead of the scales. If the movement that you obtain is small you could use a lever to make it bigger.
What property of the upholstery foam are you making use of?
Can you think of two advantages and two disadvantages of using foam in this way?
What could you use that would be more reliable.

Moments project (2)

Design a weighing machine which will weigh objects up to 5 kg which is accurate to the nearest 250 g.

Five bags of sugar weigh 5 kg. The small container holds 250 g

THE WINDLASS

Levers are also used to make it easier when pulling a rope. The diagram shows a windlass.

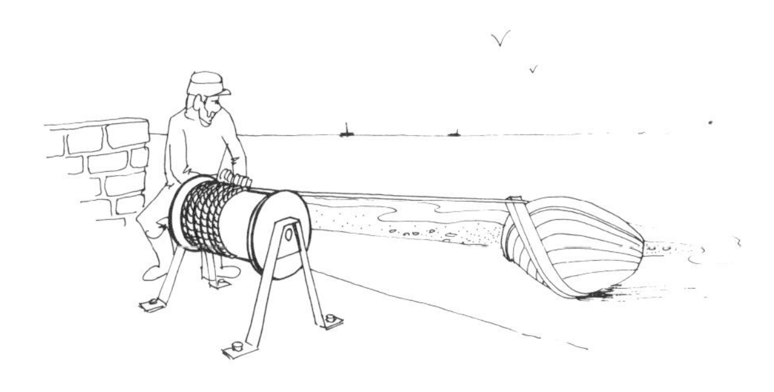

A windlass being used to pull a boat up the beach

By turning the handle the rope is wound onto the drum.

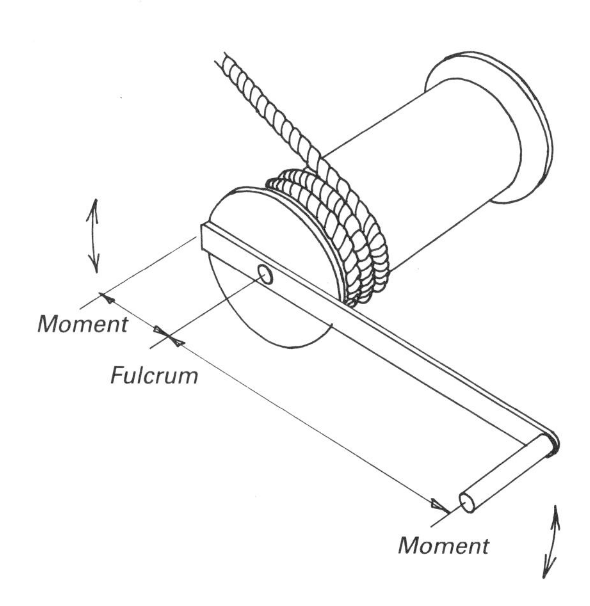

Because the winding handle is long it is easier to wind the rope than to pull it by hand.

Investigating the windlass

Meccano sets usually have a handle (a crank) which can be used to make a windlass. Make a simple frame similar to the one illustrated and find what the maximum weight is that you can wind up. Take care not to let go of the handle as it may fly back and hurt you. Using fishing line or very strong string and fasten it to the axle as shown in the diagram.

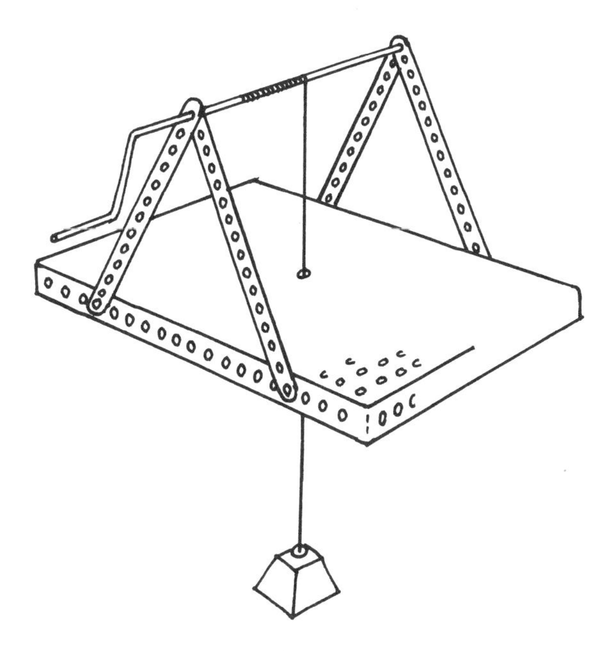

The diagram shows the line passing through a hole in the base plate. Can you see a disadvantage to this method? How could you improve it?

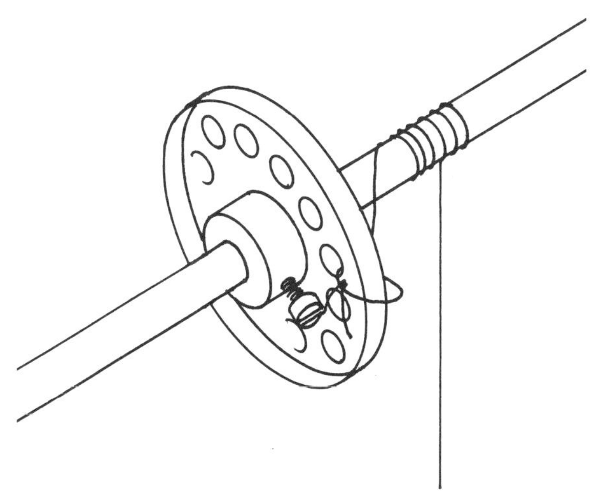

Tying the line to a gear or pulley will stop it from slipping around the axle.

Try lifting the weight by holding the string. Which is the easier? How could you improve the design? What alterations would you have to make to the length of the handle?

The windlass is used on break-down lorries and for pulling heavy loads onto trailers. A ratchet prevents the handle from flying back when you stop winding. The ratchet can sometimes be reversed so that you can unwind as well as wind in a controlled way.

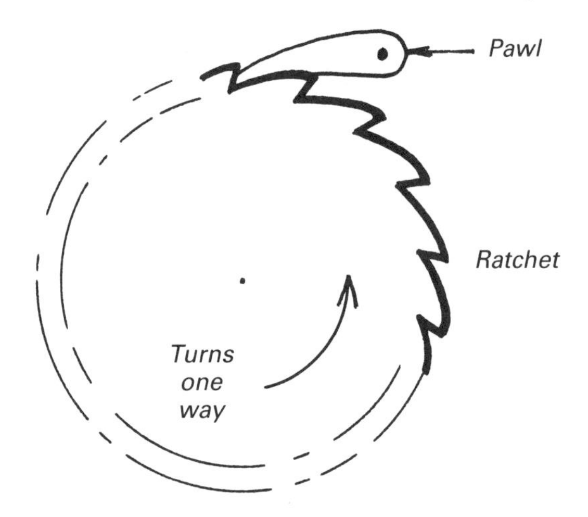

As the ratchet turns the pawl clicks from one tooth to the next but as soon as the ratchet tries to turn back the pawl locks (stops) it.

The ratchet shown could stop in many different places. Design a ratchet and pawl for your windlass. Provided that your design stops the windlass from turning you do not have to make it stop in as many places as the one shown. Design and make a ratchet which can be reversed so that it will lock in either direction.

Some clocks have a ratchet which clicks when they are wound by hand. What is the difference between a clock ratchet and the one shown?

Most bicycles have a ratchet so that you do not have to pedal when going downhill unless you want to.

To find out more about the ratchet it is easiest to turn the bicycle upside-down. If you do, make sure that you have permission first. There are many ways that you could hurt yourself doing this and it is better if an adult helps you. Make sure that you will not damage the bicycle.

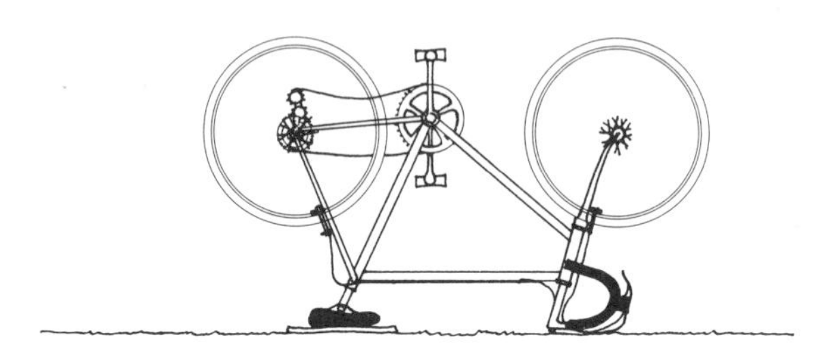

Turn the front wheel first and listen for any sounds. Turn it in both directions. Does the front wheel move freely in both directions? Now turn the back wheel and listen for any sounds. Beware of the pedals turning at the same time as the wheel. Make sure that your fingers are clear of the chain.

You will probably hear a ticking or clicking sound especially if you turn the wheel by the pedals and then hold the pedal still while the wheel rotates.

Can you explain why there is no clicking when the front wheel is turned?

Without taking anything apart, can you tell how many 'teeth' there are on the ratchet?

Can you explain why there might be more than one pawl in contact with the ratchet?

If you look after your bicycle very well, the clicking may be very hard to hear. Look at page 73 to understand why.

Transmitting power

When we wind a handle which works a windlass the power from our hand is being transferred through a lever to the windlass and along the rope which pulls the weight. This is called the transmission of power. Electric and clockwork motors produce rotary motion. Look at page 51 for examples of other sources of rotary motion. It is not always possible to have all the things that you want to turn on the same shaft.
When a crank turns in a circle it produces rotary motion.

PULLEYS

Pulleys and belts can be used to transmit power from one shaft to another.

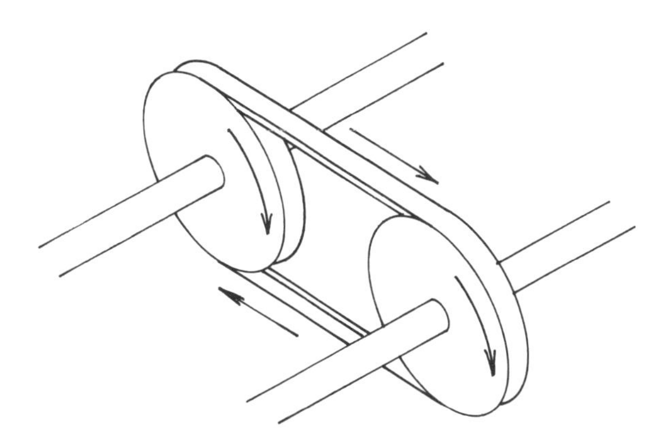

When one pulley drives another they both turn in the same direction

Machines in school workshops are often driven by belts which are like very strong elastic bands transmitting power from one pulley to another. Do not remove the covers from a machine without permission.

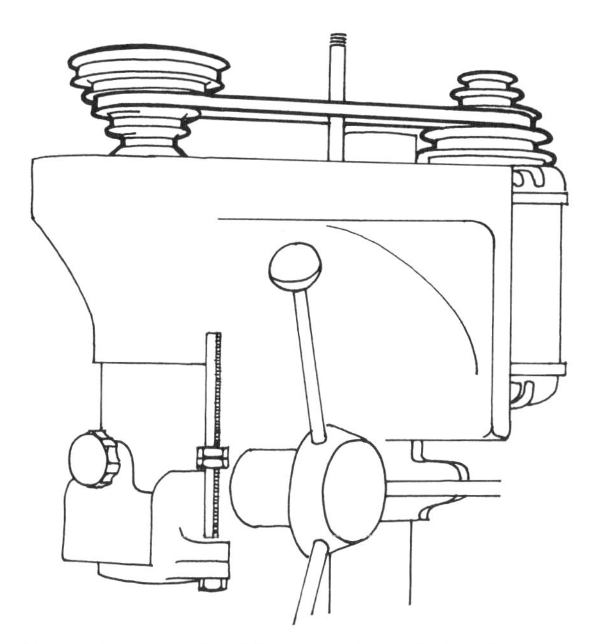

Power is transmitted from one 'cone' of pulleys to the other using a belt

An advantage of the belt and pulley is that if there is an accident there is a *slight* chance that the belt may slip which would prevent too much damage or injury. Do not rely on the belt slipping—even a loose belt will transmit enough power to hurt you seriously.

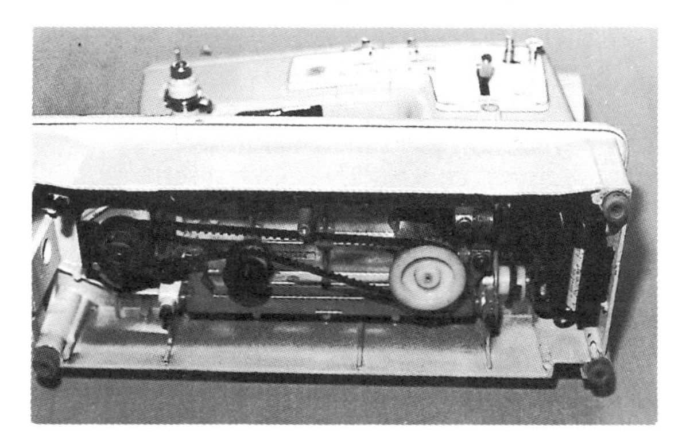

A toothed timing belt in a sewing machine

If a belt is too loose it will not transmit all the power from one pulley to the other.
Can a belt be too tight?
Using the pulleys and a belt from a construction kit, make a system so that the distance between two shafts can be increased. When the belt is stretched do you need more or less force to turn the pulleys? How can you explain the result?

Belts are used to drive the electrical generator of a car engine. The fan which helps to keep the engine cool is also sometimes driven by a belt. An adult should be present if you want to look inside the bonnet of a car.

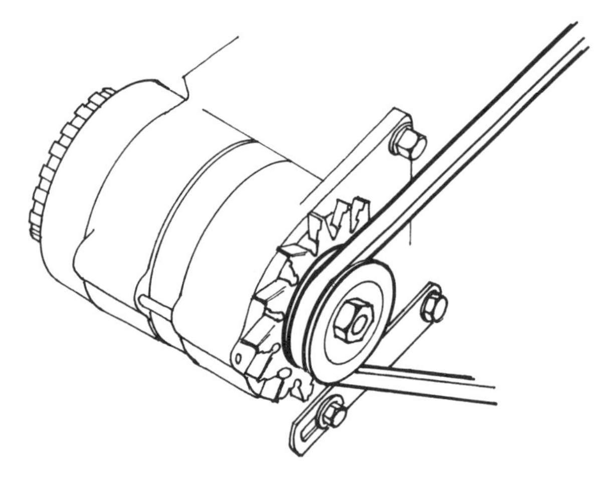

The belt drive to a car's electrical generator.

Investigating pulleys

Using Technical LEGO® build a system of two pulleys, both of the same size, with an elastic band transmitting the power. When you turn one of the pulleys you are putting power into the machine. This pulley is called the input pulley. The other pulley is called the output pulley.

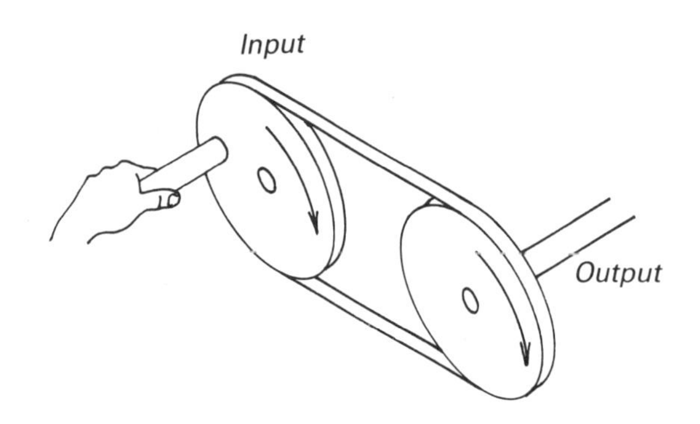

When the input pulley turns once the output pulley turns once. A way of writing this is '1 : 1' which is read as 'one-to-one'. The symbol (:) means 'ratio'—a way of comparing things. Maps usually have a scale which is expressed as a ratio.

A hand drill like the one shown has an input and an output. How many turns does the drill bit make for every single turn of the handle?

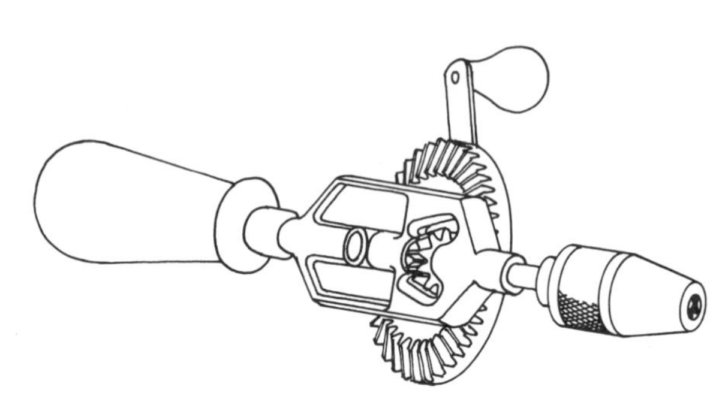

A hand drill

What is the ratio of the output to the input? Can you think of any other examples of ratios?

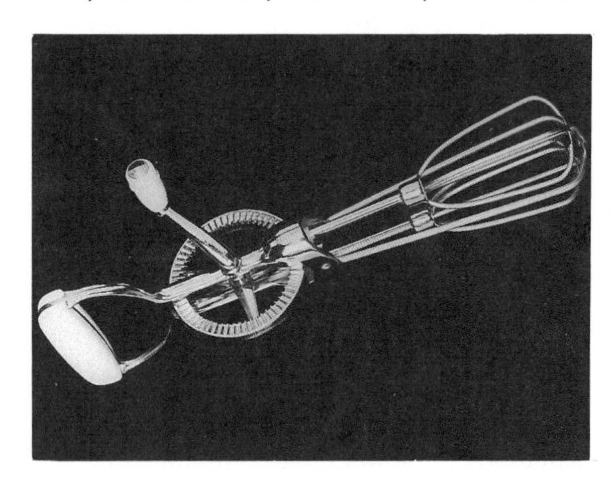

A rotary egg whisk

Build a system of pulleys using a 36 mm diameter input pulley and a 24 mm diameter output pulley. When you turn the input pulley once, how many times does the output pulley make?

Write these numbers as the ratio 1 : 1.5. It is more usual to use whole numbers which would make the ratio 2 : 3. What is the ratio of a system which has a 60 mm diameter input pulley and a 20 mm diameter output pulley?

The ratio can be worked out by using the formula:

$$\text{ratio} = \frac{\text{size of output pulley}}{\text{size of input pulley}}$$

In the example given above:

$$\text{ratio} = \frac{20}{60} = \frac{1}{3}$$
$$= 1{:}3$$

What is the ratio if you use the same size pulleys by making the 20 mm diameter pulley the input?

When the output or driven pulley is smaller than the input or driven pulley, the output pulley will turn faster than the input.

When you make a pulley system to your own design you will be limited by the size of elastic band that you can use as a belt. You will also have to think about how the pulley is going to be fastened onto the shaft.

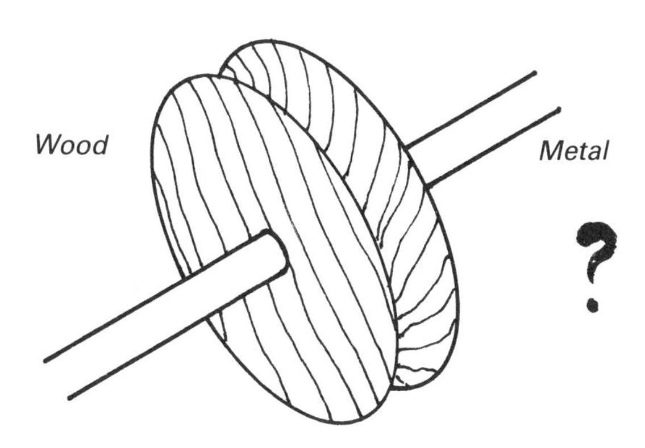

Can you think of ways of making pulleys? Simple pulleys can be made by glueing together three disks of thin plywood or thick card.

Sometimes it is easier to think of a system which solves a problem as being a black box. This does not mean that everything has to fit into a box. It means that when you have a problem to solve you may know what the input is going to be and also what output you want but it is the part in the middle that has to be solved. The part in the middle can be drawn as a black box.

All kinds of unsolved problems can be described in this way. For instance:

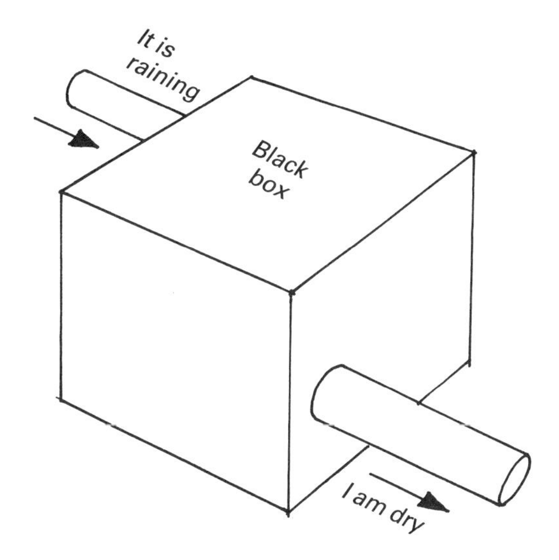

A number of things might fit into the box such as 'I have an umbrella', 'I am indoors' or 'I have oily feathers'. How could you fill the black box in the next example?

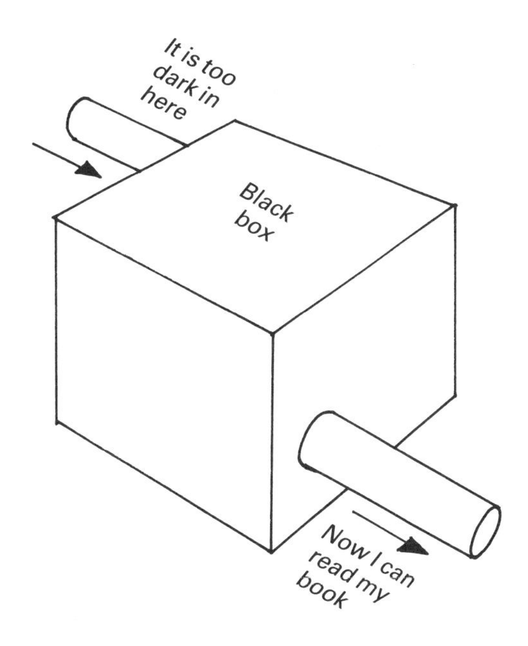

Try to think of some unusual ideas as well as 'I shall switch on the light'.

Practical problems can be drawn like this:

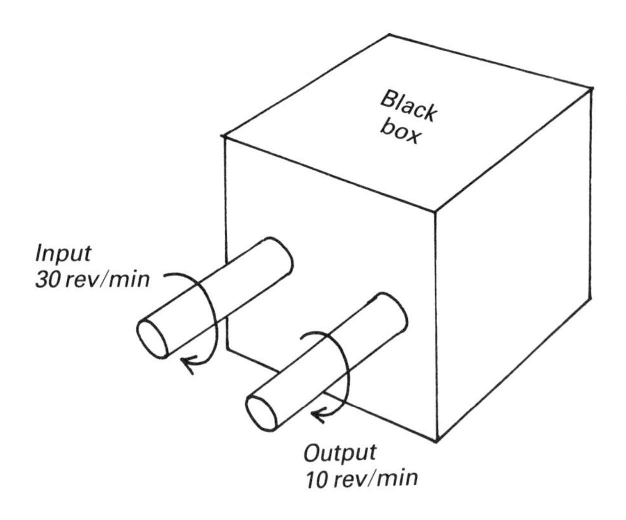

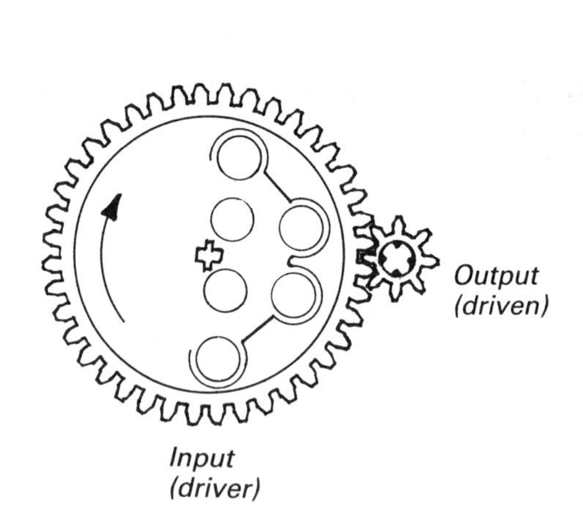

Gears can be used to solve the black box problem shown in the diagram.

Can you remember what RPM means? See page 52 if you cannot. Your solutions or answers do not have to fit into a box but the shafts do have to point in the same direction as shown in the diagram. Make your solution (and test it) using Technical LEGO®.

GEARS

Pulley and belt systems are very useful but there are times when the belt can slip and this causes problems. Gears transmit power without slipping.
Use Technical LEGO® to make a gear system where the driver gear has 40 teeth and the driven gear has 8 teeth. The teeth of the gears must be the same type and size for them to fit or 'mesh'. When the driver gear turns clockwise which way does the driven gear turn? How can you work out the ratio of one gear to the other?
You could measure the diameter of the gear just as you did for the pulley but there is another way that can be used for gears.
You can find the gear ratio of a single speed bicycle in the same way.

How would you find the gear ratios of a 3 or 4 or 10 speed bicycle?

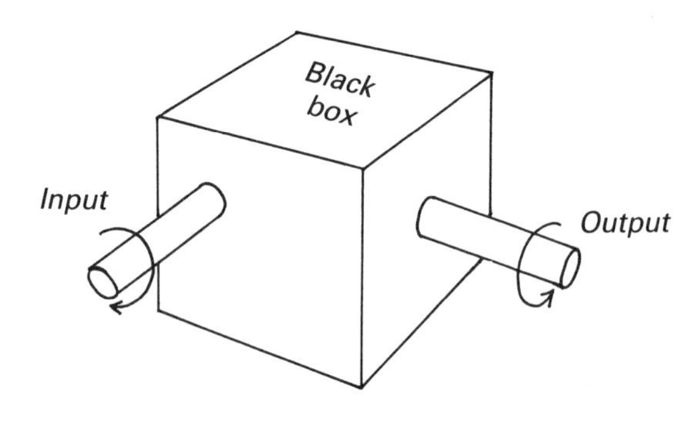

Solve the problem using a construction kit.
The gears and pulleys that you have used so far have all worked in the same plane. This means that they have all been in line with one another. When you want to transmit power through 90° you will find the bevel or mitre gears useful.
Draw a diagram to show how you could also solve the same problem using only pulleys.
You may find that isometric grid paper is particularly useful for this. Why?

By using a large driver gear and a smaller driven gear it is possible to make the output faster than the input.

Investigating gears

Use any of the straight gears in a Technical LEGO® set and make a system of gears (gear train) that produces the largest number of turns from one input turn. You will have to have more than one gear on some of the shafts. Put a marker on the output shaft to help you count the turns.

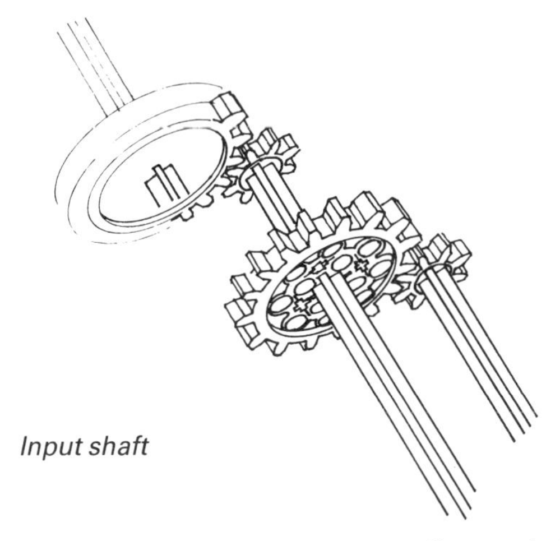

Input shaft

Output shaft

You should easily get a ratio of 1 : 20 and perhaps even more. When you add more gears is it harder or easier to turn the handle on the input gear? Why do you think this is? Read page 48 to remind you.

When the shafts in your gear train rub on the holes in the blocks (the bearings) friction wastes some of the energy that you want to be used for turning gears. You may find that even though you can build a gear train which should give you a higher output, friction makes it impossible to turn the input gear.

Friction can be reduced by using a lubricant such as oil. The oil becomes trapped between the moving parts and helps to stop them from rubbing so hard against each other. If you rub your hands together whilst they are soapy they will not get as hot as they would when dry. In a car engine the heat from all the moving parts is reduced by the oil which is circulated (moved around) by a pump. Without using oil to reduce friction a car's engine would rapidly overheat and wear out.

Do not lubricate (oil) the parts in a construction kit unless told to do so by your teacher.

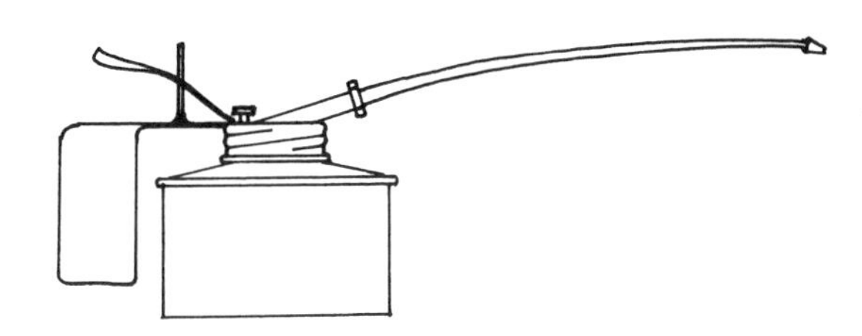

An oil can

Investigating oil

Look at a cycle and find the parts which look like those illustrated.

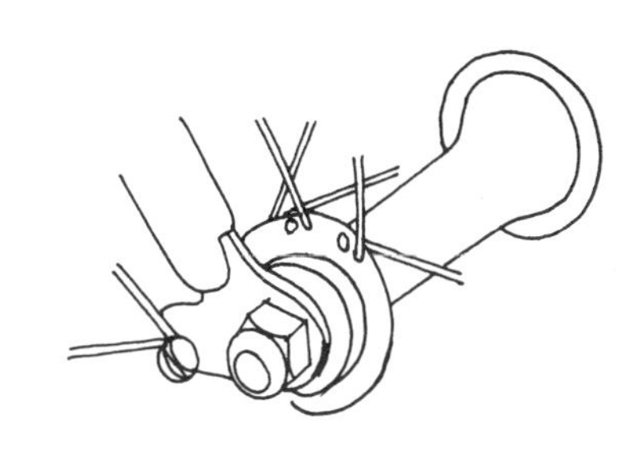

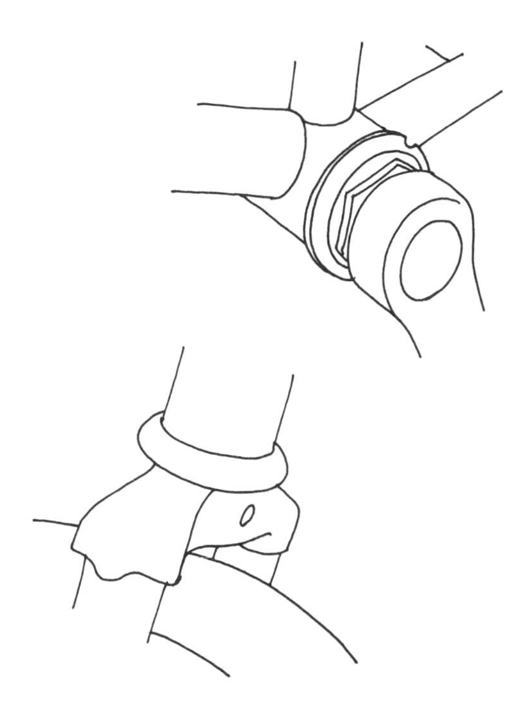

Oil can be bought in different grades (thicknesses). Find out the grade(s) of car engine oil and if possible find out the grade of oil used for cycles and other light machines.

What also does oil do besides reducing friction? Look at page 33. Describe the places where oil could be used to coat ferrous metal on a cycle. When oiling a cycle be sure that you do not oil the rims of the wheels where the brake blocks press. Friction is very important if the brakes are to work properly. What materials rub together to make the cycle's brakes work? Do cars have the same materials in their brakes?

Car brake drum

The machines that have been described so far all have parts which move in a circle (rotary motion). Parts can also move in a straight line (linear motion). An example of linear motion is the movement of the handles on a table-football machine.

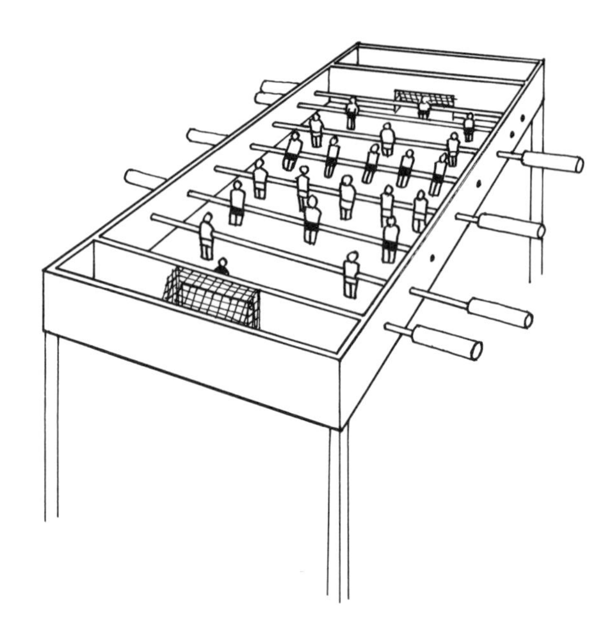

Some machines have a rotary input and a linear output. If this was shown as a black box it would look like this:

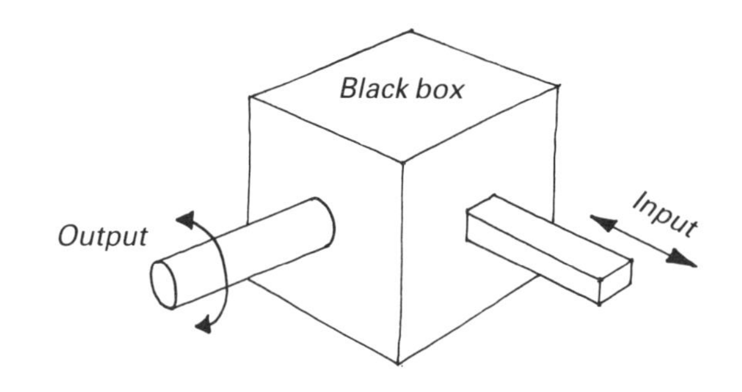

A sewing machine powered by an electric motor has a needle which moves up and down in a straight line.

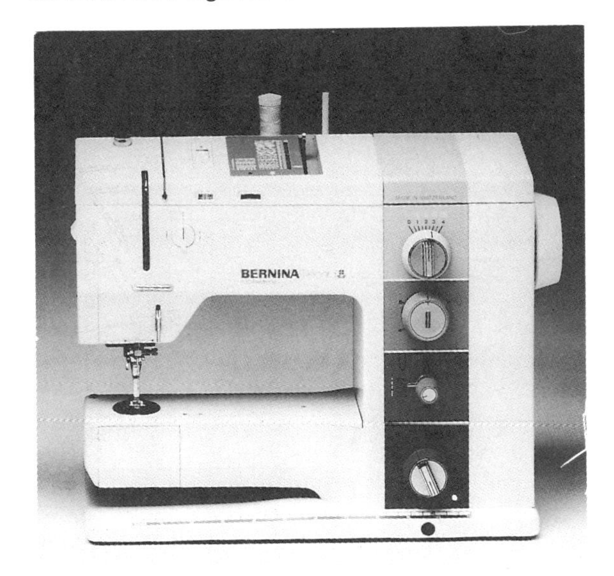

Investigating linear and rotary motion

The illustration shows how a rotary motion can be converted into a linear motion.

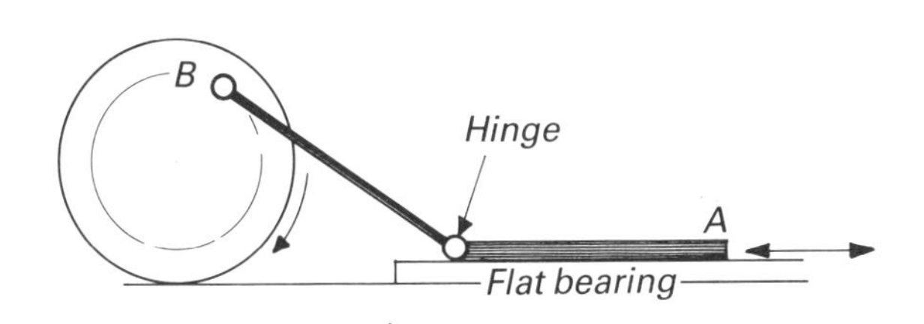

Does it make any difference if 'B' rotates anti-clockwise?

Build a working model using Technical LEGO®. What happens to the distance moved by A when B is moved closer to the centre of the wheel?

Can you design and make a solution to this problem?

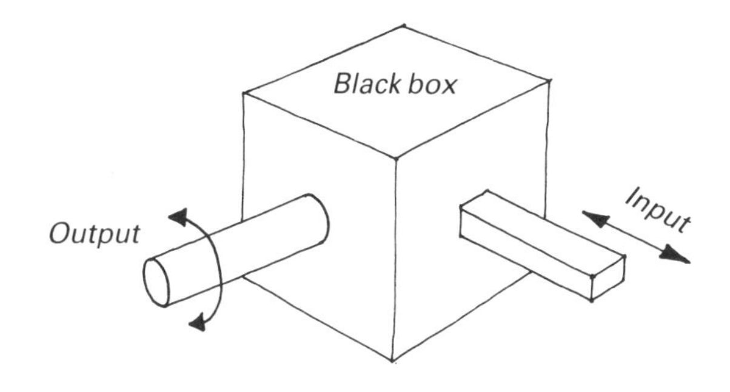

You may find that parts which look like this could be useful.

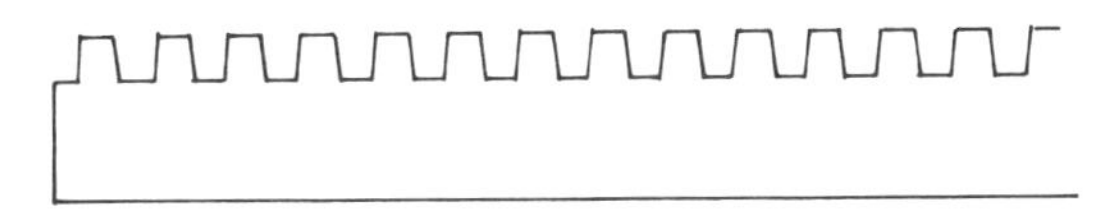

A cam

Rotary motion can also be converted to linear motion by use of a cam. There are many different types of cam but the illustration shows one of the easiest to understand.

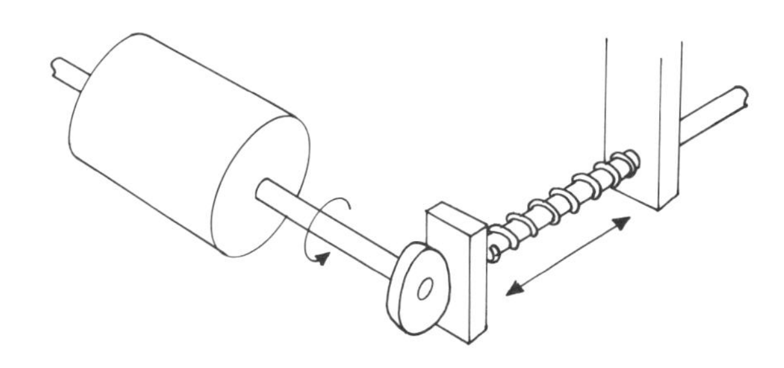

The cam on the driveshaft of the motor pushes against the follower and compresses the spring

Cams are used to open and close the valves in a motor-car engine.

There is a cam in some Fischer Technik sets which will help you to understand how they work. A simpler way of using this idea is to make an eccentric. An eccentric is a circular part which is turned so that it does not revolve around its centre. Because the end of the axle that is being lifted is moving in a curve it is called an oscillating motion (like the pendulum of a clock).

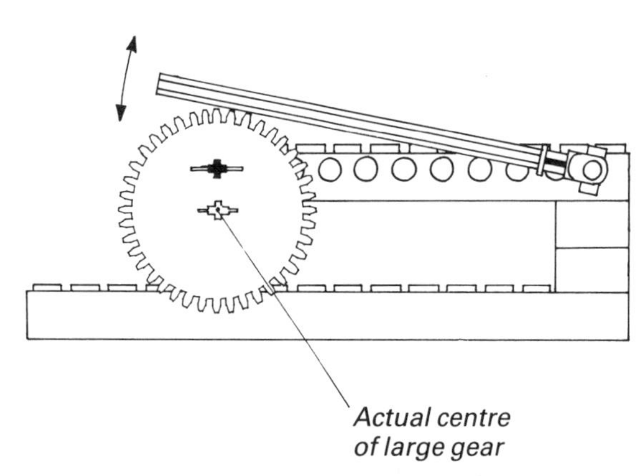

An eccentric made in Technical LEGO®

Using Technical LEGO® design and make a system which uses an eccentric and has a linear output.
When the linear output goes backwards and forwards it is called a reciprocating motion. The needle of a sewing machine has a reciprocating motion.

Investigating mechanisms

Look at p. 77 and using a construction kit, solve as many of these black box problems as possible. When you have solved each one draw a diagram of it. Try to think of at least two different solutions for each problem and record your ideas.
Can you think of an example of where each mechanism is used?

For example:

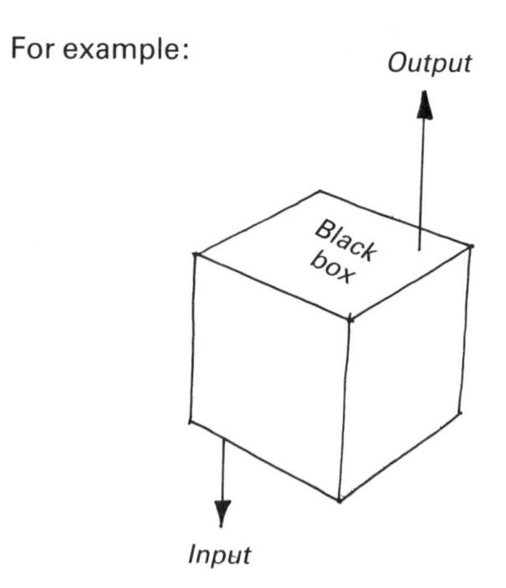

could be solved by:

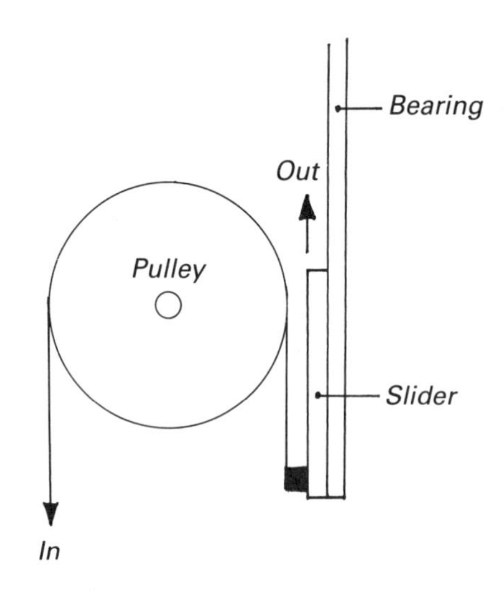

or:

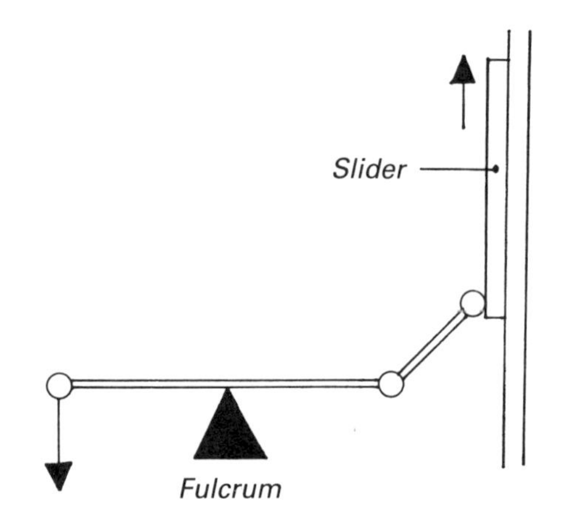

Black box problem

The arrow heads are very important.

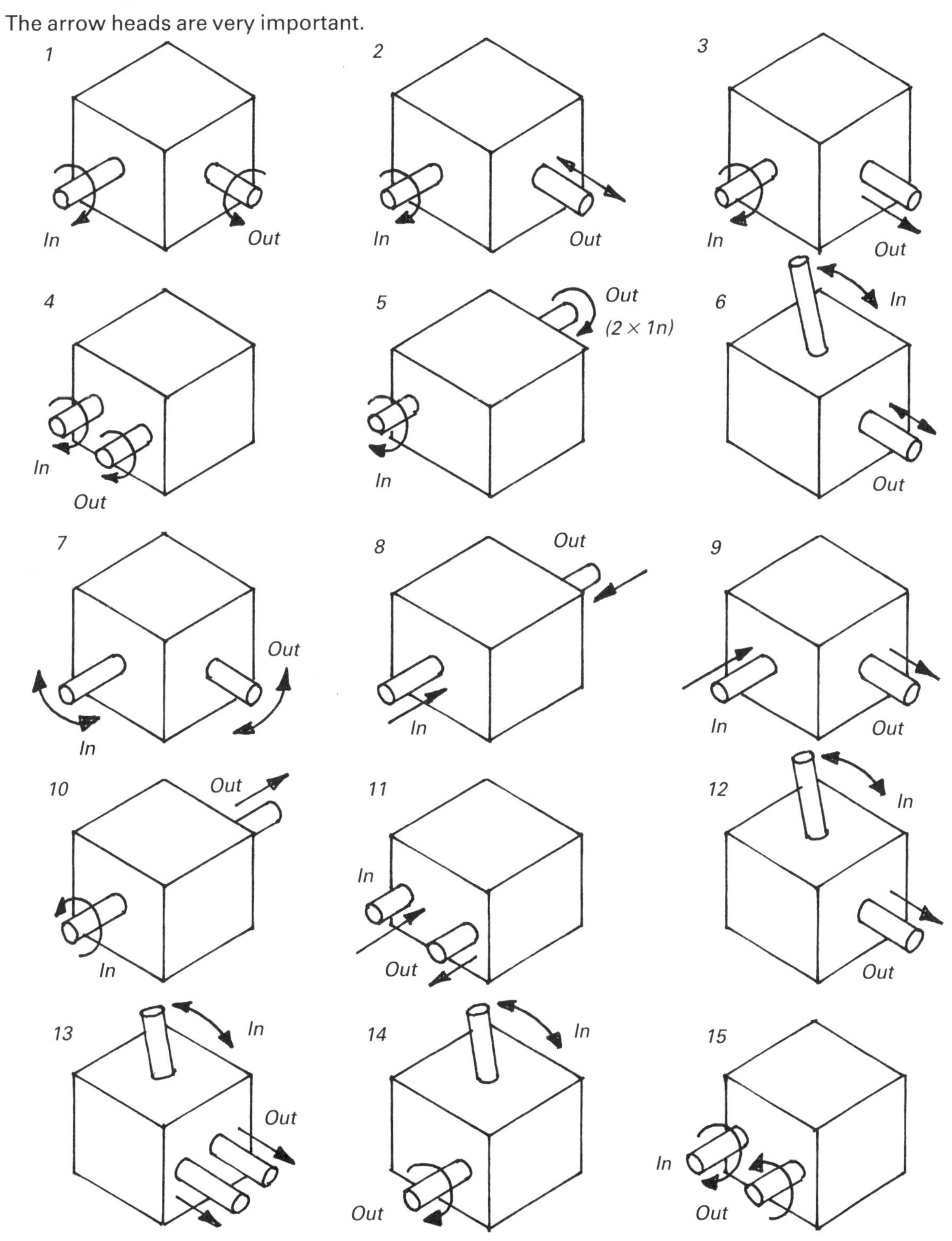

Make up some of your own and draw them on isometric grid paper.

Mechanisms project (1)

The illustration shows an idea for using wind power to drive a bird scarer or a garden novelty (toy).

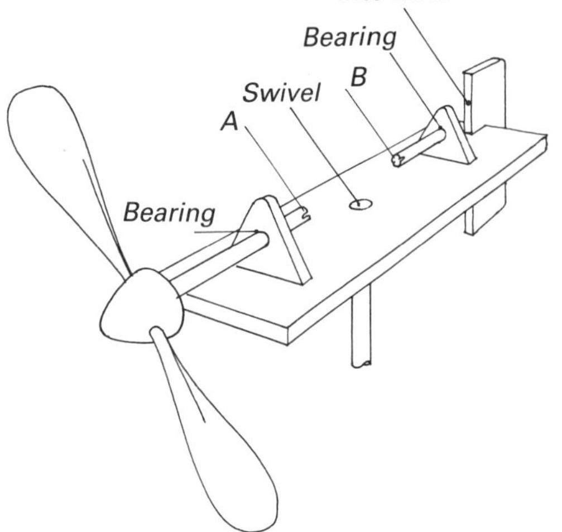

You will need to make sure that the engery wasted by friction in the bearings is kept to a minimum. Wooden parts which rub together will produce too much friction.
Which materials make good bearings?
Use a diagram to show how a crank mechanism could be fitted between A and B.

The crank will give rotary motion. Draw two diagrams to show how this motion may be used. How could the rotary motion be converted into linear motion? Other mechanisms that could fit between A and B might be a cam or an eccentric. (See page 76.) Show two different designs for using the motion produced by them.

Mechanisms project (2)

Design a toy that would be suitable for a four-year-old child to play with. Your design must include a light which flashes and one part which moves another.

Write a list of details which you think may be important. Copy the points given here and add at least four or your own.

(i) Because toys are often knocked about, the surface treatment must not flake easily and parts must not break off easily.
(ii) Electrical connections, batteries and wires must not be accessible to the child when the toy is in use.
(iii) The toy must not have sharp edges or be dangerous.

CONTROLLING ELECTRICITY

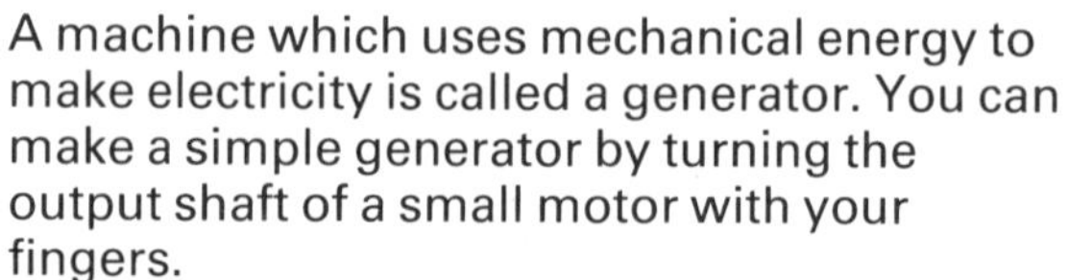

A machine which uses mechanical energy to make electricity is called a generator. You can make a simple generator by turning the output shaft of a small motor with your fingers.
Connect a d.c. voltmeter to the terminals of the motor and you should get a reading of about 0.25 V. Be careful when doing this —if the meter's needle starts to move backwards then turn the shaft in the opposite direction.

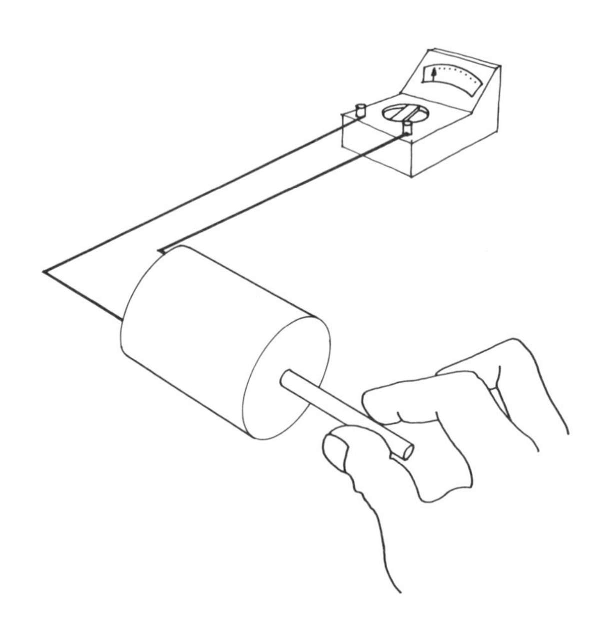

Electricity is produced when a coil of wire is moved in a magnetic field. Electricity is a movement of electrons from one atom to the next in a conductor. Refer to page 26 if you cannot remember what a conductor is. The electrons do not really flow like water but we often describe their movement as the 'current'. Electrons are more like a line of dominoes.

When the first one is knocked over it knocks the next one over and so on until the end of the line.

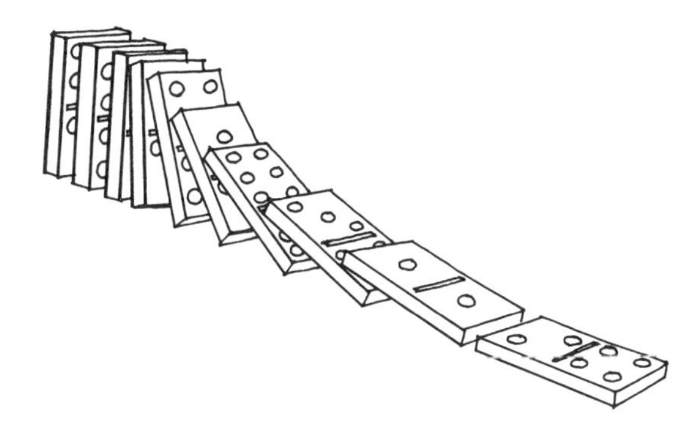

Drawing all electrical circuits this way would be too complicated. Symbols are simple pictures which help us to draw circuits quickly. The circuit could be drawn in the following way using symbols. The switch is shown open or in the off position.

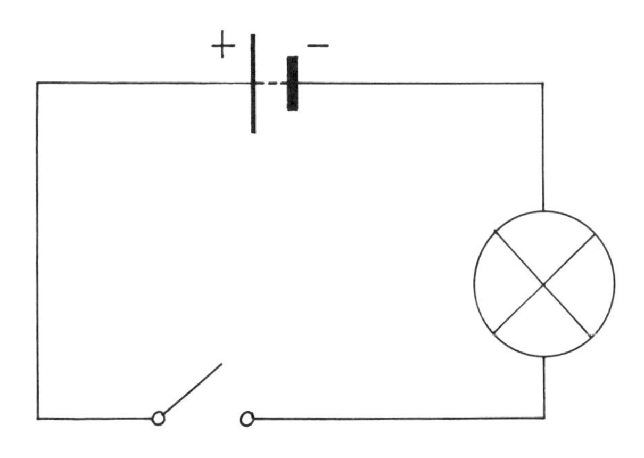

If you connect a wire from one end of a cell to the other the electrons flow from the negative end to the positive end. However, it is not a good idea to do this—the electrons will rapidly stop moving and the cell will become 'flat'.

By controlling the movement of electrons we can make use of electrical energy. The illustration shows a simple circuit. The switch is on and the bulb lights up.

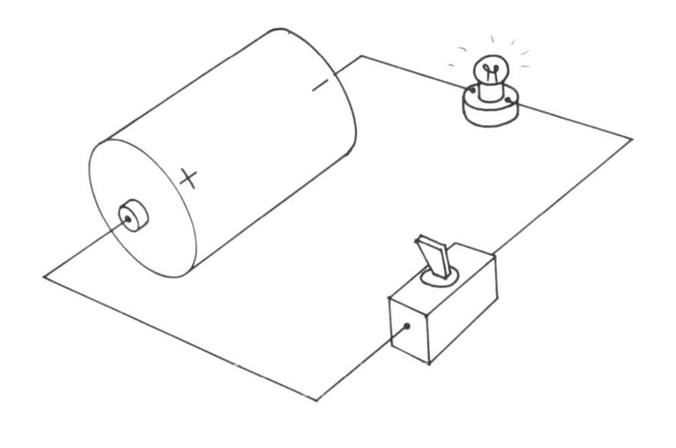

The Service Section of this book shows you ways of connecting components (parts) together either temporarily or permanently.

When the switch is open the electrons cannot jump across the gap because the air does not allow them to flow easily through it. Air, plastics materials and dry wood have a high resistance to electricity. What name is given to a material which does not conduct electricity? See page 26 if you cannot remember.

Ordinary water conducts electricity. It is very dangerous to use electrical appliances near taps, which is why there should be no electrical mains sockets in a bathroom.

How do you switch on the bathroom light?

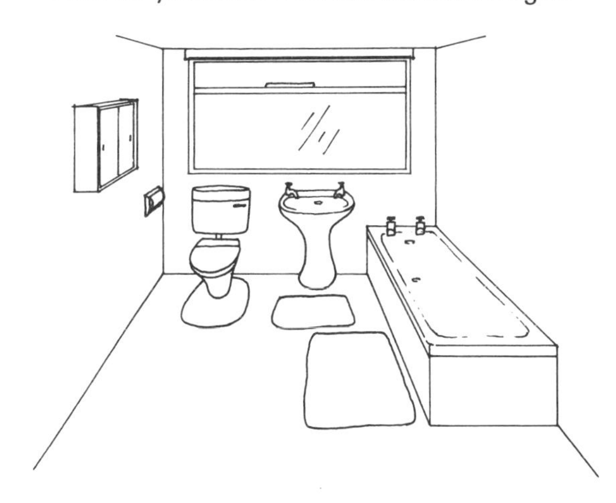

Investigating electrical flow

Build the circuit illustrated. Use a 6 volt battery (group of cells) and leave out the switch.

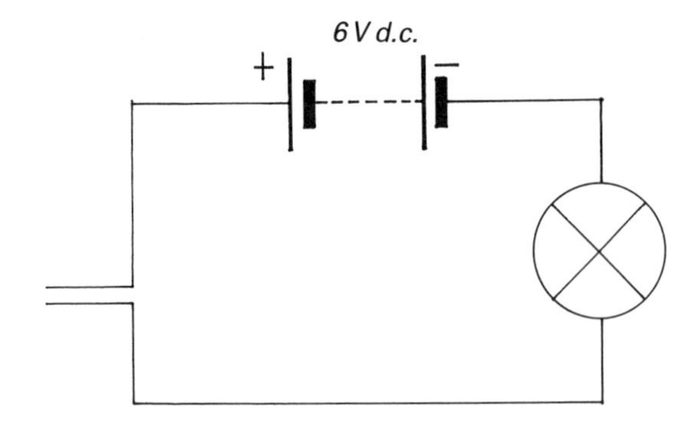

Keeping the wires apart dip them into a dish of water. Does the bulb light up?
The battery cannot force enough electrons through the water to light the bulb. By using other electronic components we can make the tiny flow of electrons cause a light to come on. The electricity from a 6 volt battery is not enough to hurt you if you should touch the wires but do not attempt this with a more powerful battery. Never get anything wet which is plugged into a wall socket.

Components

The circuit will use the following components: a battery; insulated wires; resistors; a light-emitting diode; and a transistor.

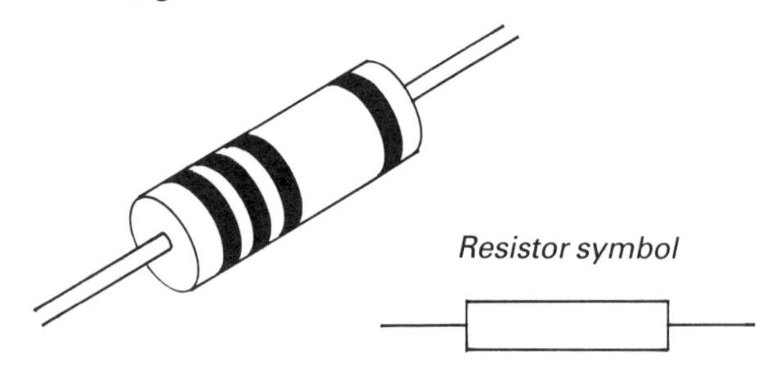
Resistor symbol

The illustration shows the resistors. They have coloured bands which tell us how much they will slow down the flow of electrons (their resistance). Resistance is measured in ohms (Ω). The coloured bands which tell us the value of the resistor are in a group of three at one end.

For this circuit you can ignore the band on its own at the other end which is usually gold or silver. Hold the resistor so that it has the three bands to your left and look at the colours. Resistors values can be shown as three numbers

First band	Second band	Third band
(first number)	(second number)	(number of noughts)

The numbers for the colours are;

Black	zero	or	no	noughts
Brown	one	or	one	nought
Red	two	or	two	noughts
Orange	three	or	three	noughts
Yellow	four	or	four	noughts
Green	five	or	five	noughts
Blue	six	or	six	noughts
Violet	seven	or	seven	noughts
Grey	eight	or	eight	noughts
White	nine	or	nine	noughts

The resistors for this project have the following values:

orange, orange, red is 3,300 Ω
brown, black, red is 1,000 Ω
yellow, violet, brown is 470 Ω

Resistors can be connected into a circuit either way around.

The diode

Instead of using a bulb the light will be a light-emitting diode. A diode only allows electrons to flow through it in one direction rather like a turnstile which only allows people to pass through it in one direction. The symbol for a diode is

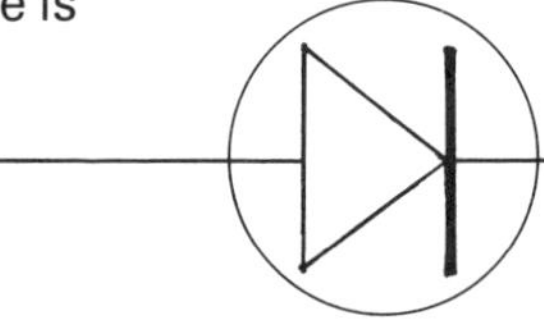

The symbol for a diode which gives off (emits) light is

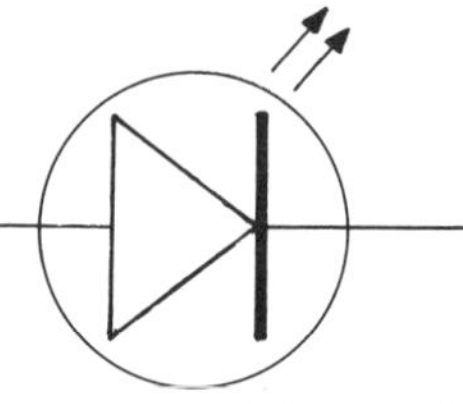

A light-emitting diode (often referred to by its initials LED) needs less power to make it light up than a bulb does. An LED is shown in the illustration.

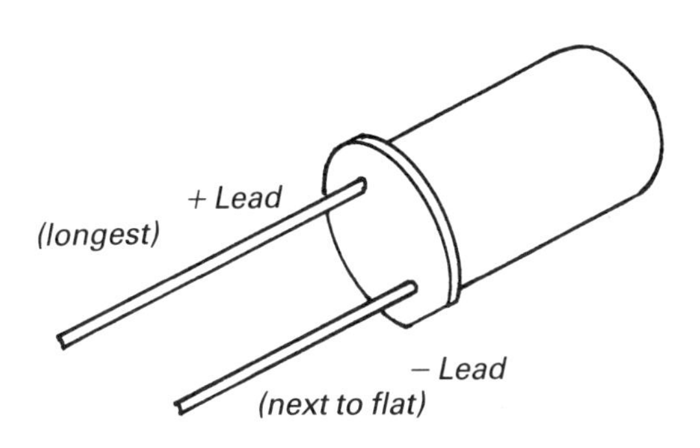

Only when the arrow of the symbol points towards the negative (–) side of the circuit will the current flow. When putting a diode into a circuit make sure that it is the correct way around and that it is protected by a resistor in series.

The transistor

The transistor has three leads (wires) and is illustrated below.

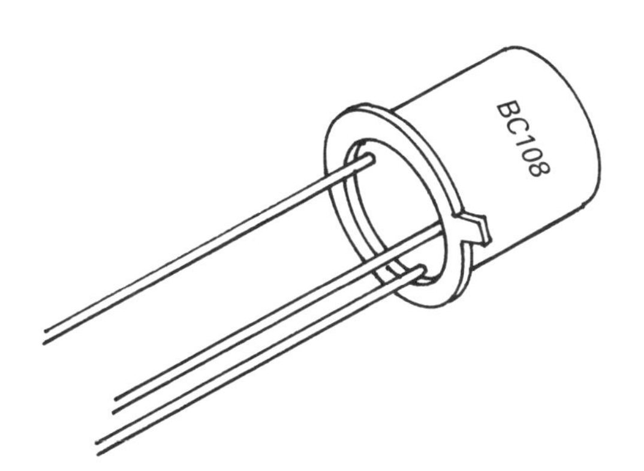

Hold the transistor so that the leads point towards you. You will notice that there is a small tab which is called the can marker. The lead next to the marker is the emitter and the others can be identified from the illustration which shows a pin-side view (one looking straight at the leads).

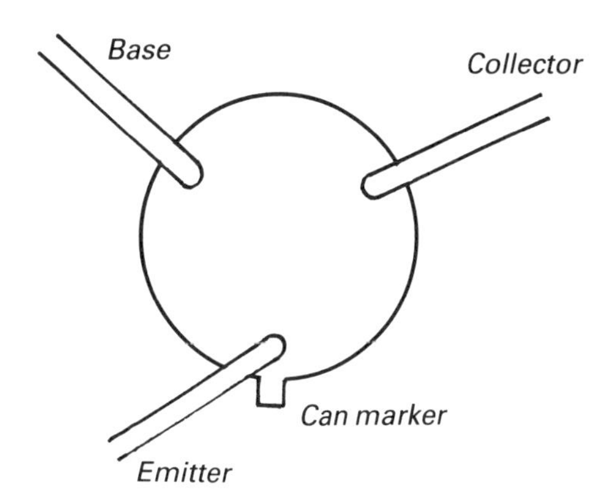

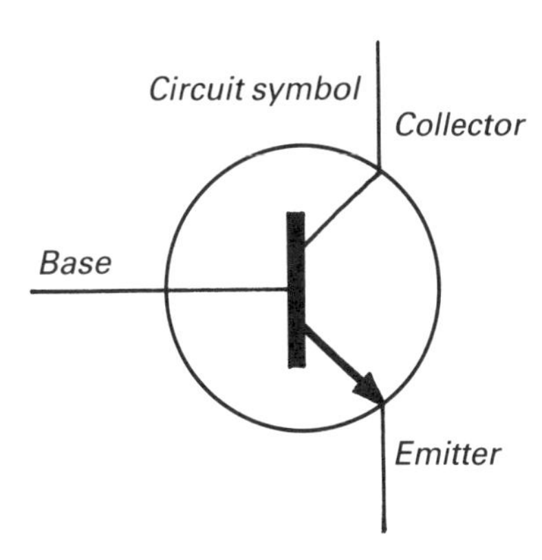

When the transistor is connected into a circuit, current will not flow from the collector to the emitter until a small current goes into the base. The tiny current which will pass through drops of moisture is enough to make the transistor switch on the LED. The resistors ensure that the current in the circuit is not too much for the transistor or the LED.
The components must be arranged so that the electrons will do some work for us on their way through the conductors. The components must be drawn in their proper positions before the circuit can be built up. This drawing is called a circuit diagram and it makes use of symbols like the ones already shown.
By writing down a list of components from a circuit diagram you can check that you have them all before you start.

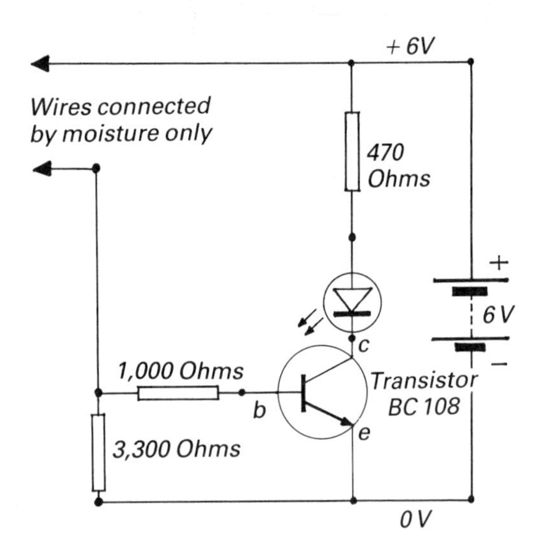

The illustration shows a moisture sensor circuit. When you build the circuit the transistor can be put in upside-down to make it easier to connect but if you are using a breadboard (see the service station) this may be difficult. If you do mount the transistor upside-down, make sure that the leads do not touch the case.

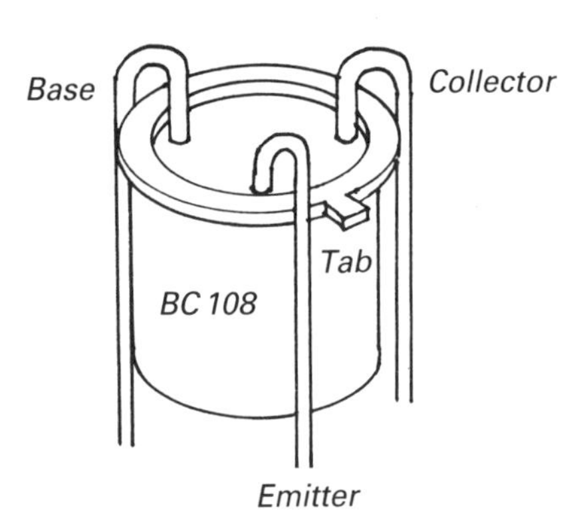

A BC 180 transistor with its leads upwards for easier identification

Be careful when bending the pins of a transistor. They are quite easily damaged.

The illustration shows the components if the transistor is used with its pins downwards.

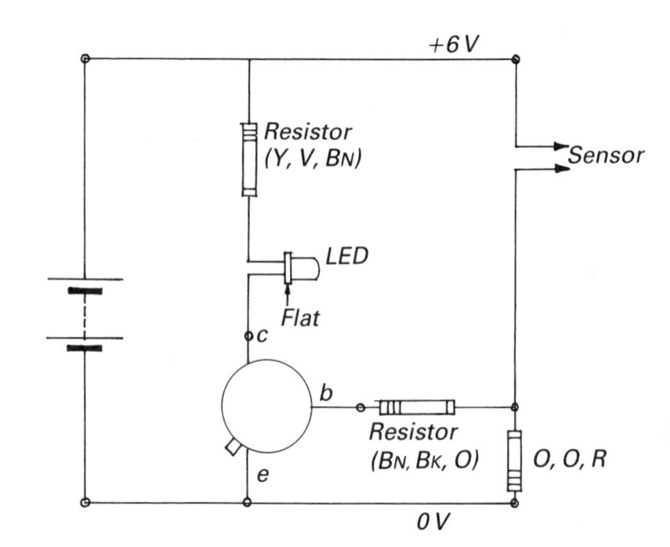

Test the circuit when complete by holding the wires with arrows onto a wet finger. The LED should light. If it does not check all the connections and also that the transistor is correctly connected. If the LED still does not light try connecting it the opposite way around. When the circuit works, design and make a better way of getting moisture to bridge the gap between the arrowed wires.

The circuit could be used to give a warning when it is raining or when a bath is full. You would have to have the sensor (the part which gets wet) at the ends of long wires and it would be better if you could design a case for the battery and circuit board.

Could this circuit be used to show when a plant pot needs watering?

How would you adapt (change) it?

When someone is telling lies their hands are supposed to sweat more. Could you use the circuit to show if this is true or false?

Design a simple circuit for each of the following, using only a battery, LED (with a protecting resistor) and wire.

(1) A steady hand game.
(2) A cartridge fuse tester.

Service Section

This section contains information which should give you practical help with your projects.

Electronics

Circuits may be built so that the components are fastened together so that they will not easily come apart. The most usual way to do this is by soldering them to one another. Soldering is a permanent method of building circuits. The opposite of permanent is temporary. A temporary method of joining components is by using a circuit board similar to the one shown below. The leads of the components are pushed into the small square holes and connections are made by strips of metal below them.
Components can easily be put into or taken out of the circuit without soldering.

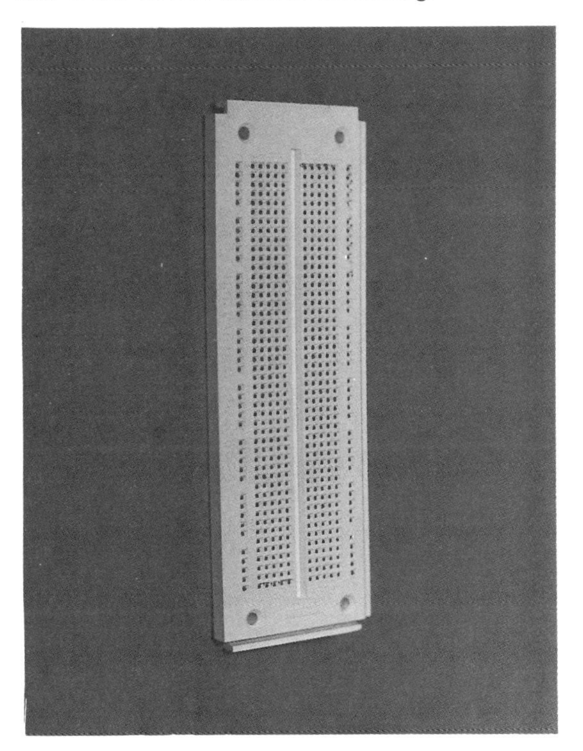

The holes along the X row are connected. The holes along the Y row are connected. X and Y are usually used for the positive(+) and the zero volts connections.

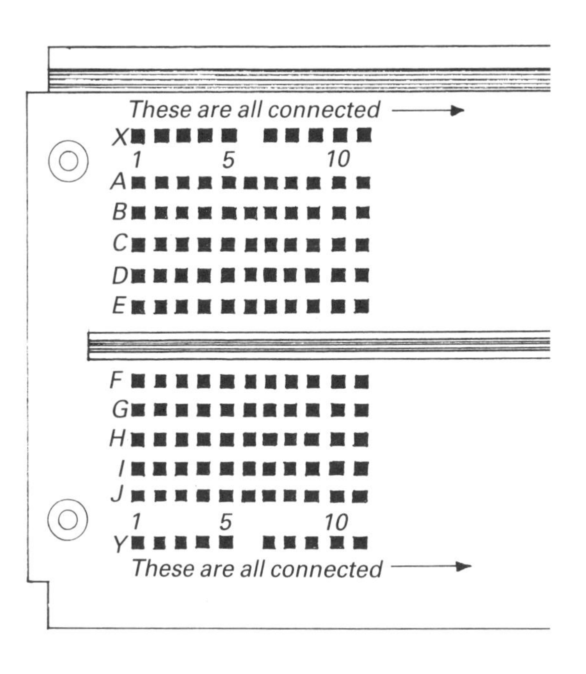

The end of insulated wires are stripped so that they can be pushed into the holes.

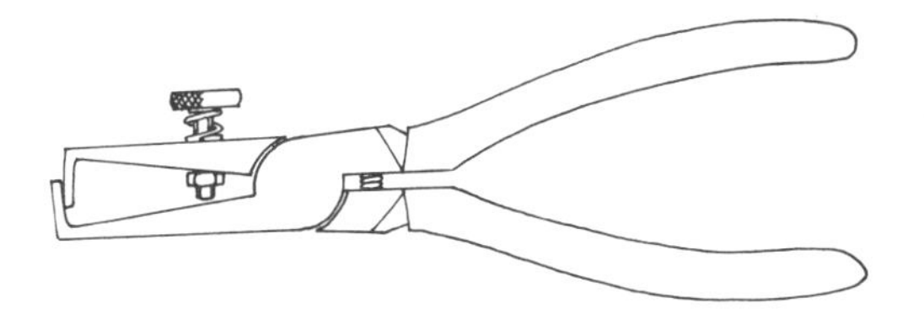

There are many different types of wire strippers. It is usually possible to adjust them so that the insulation is removed without cutting into the wire. Take care you do not 'nick' the wire when stripping off the insulation as the wire is likely to break off at that point.

Connect the positive (+) terminal of a 6 volt battery or power supply to the row of holes marked X. This is then the positive rail. Connect the negative (−) side of the battery to the Y holes. This is now the zero volts rail. Use a bulb in a bulb holder with single-strand insulated wire to find out which connections make the bulb light.

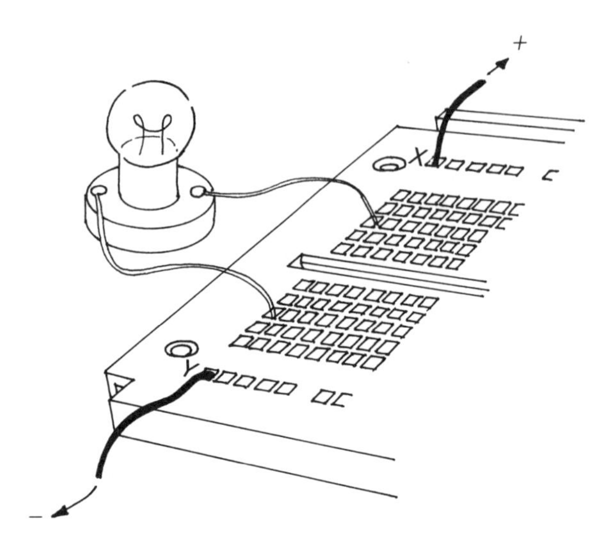

Draw a diagram of which holes you think are connected. Use two more pieces of insulated wire with stripped ends and find out if there are any other possible connections. Do not connect X to Y.

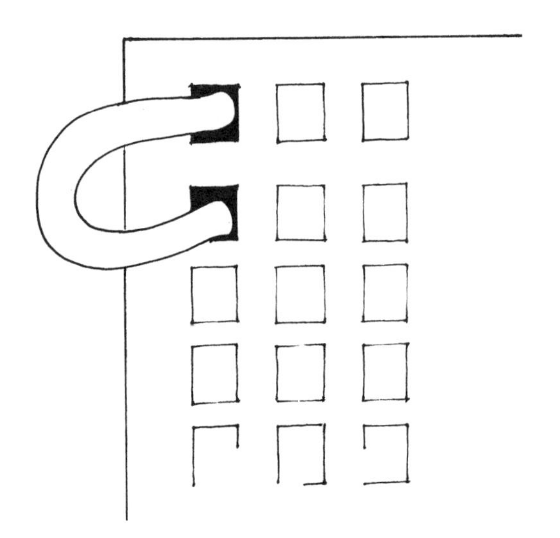

You will find that paper with printed 5 mm squares is useful when planning the layout of a circuit on this kind of board. Draw a plan of the holes with the positive (X) at the top and label the squares with the same system used on the circuit board. Draw the components with their leads ending in a square. Look at page 81. You will notice that when you look at a transistor from the top it is back-to-front with its circuit symbol. You will find it difficult to put transistors in upside down when using a temporary circuit board.

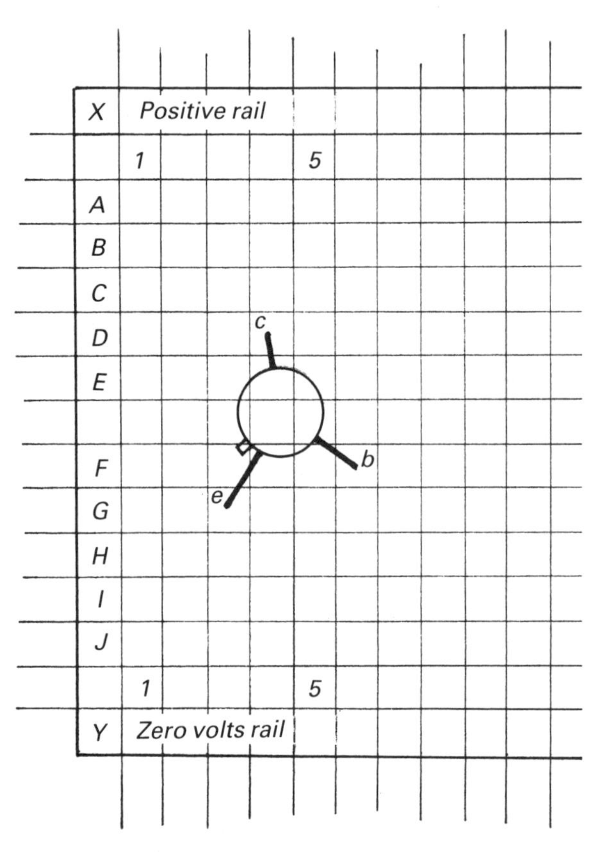

Squared paper can be used to help locate components

Copy the diagram which shows a BC 108 transistor connected to the board. These boards are often called 'breadboards.' Draw the rest of the components for the circuit on page 82. You will find it easy if you follow the illustration which shows what the actual components look like when connected.

When you want to build a circuit that is permanent there is a wide choice of systems. Temporary boards like the breadboard can be expensive but they can be used hundreds of times. One of the cheapest permanent methods of joining parts is to use steel panel pins which are driven into 6 mm thick plywood. Why do you think that plywood is better than 6 mm thick softwood? Mark out the circuit on the wood and drive in pin where connections are to be made.

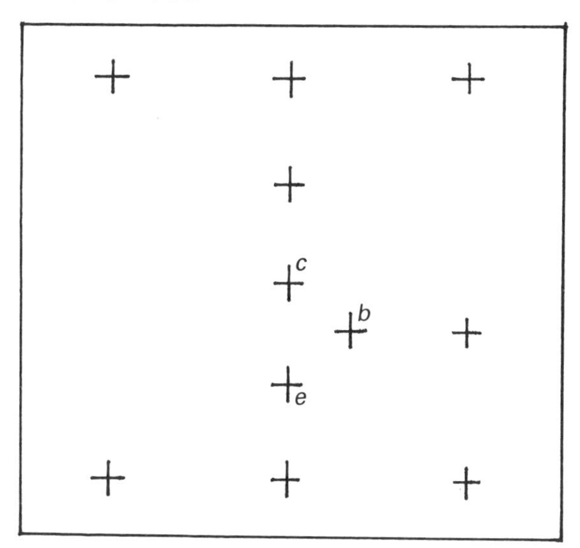

The illustration above shows the layout for the moisture sensor circuit on page 82. The panel pins are driven in where the black spots are. Assembling the parts means putting them together. Do this by wrapping the wires and leads carefully around the pins as shown below:

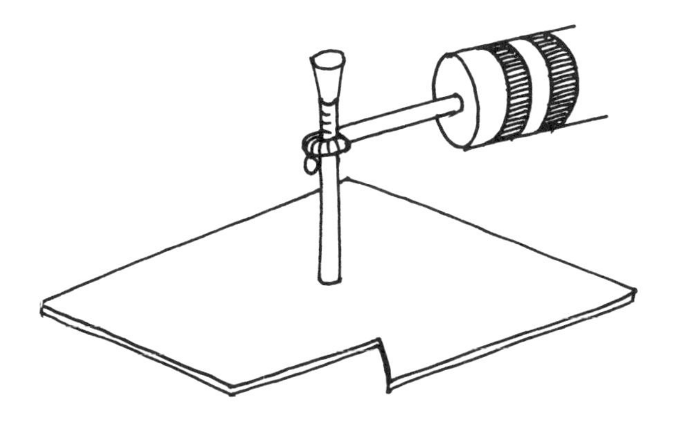

When all the parts are assembled in this way they have to be soldered together so that electricity can pass easily from one part to the next. Solder is an alloy (see page 32) which melts at a low temperature. Solder used for electrical joints has a flux down the middle of the strand which helps to clean the area around the joint. You will need a soldering iron which must be put into a special stand when not in use.

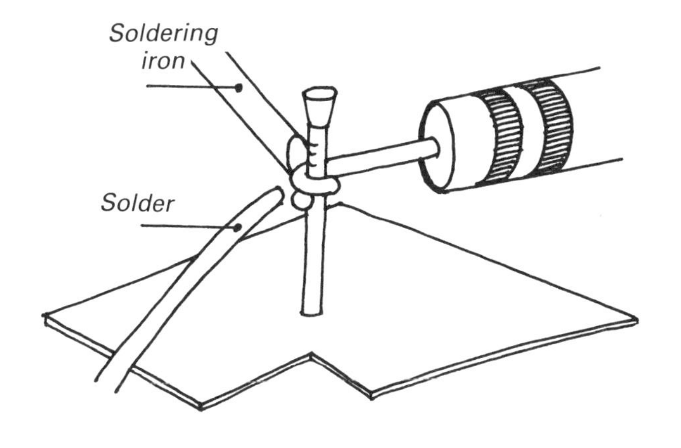

Heat the pin and the lead until the solder melts easily when touched onto them. The joint should look bright if you have made it well. Be careful not to take too long when soldering transistors—they can be damaged by being overheated.

Instead of pins and plywood you could use matrix board which has special pins fitting through a grid of accurately-made holes.

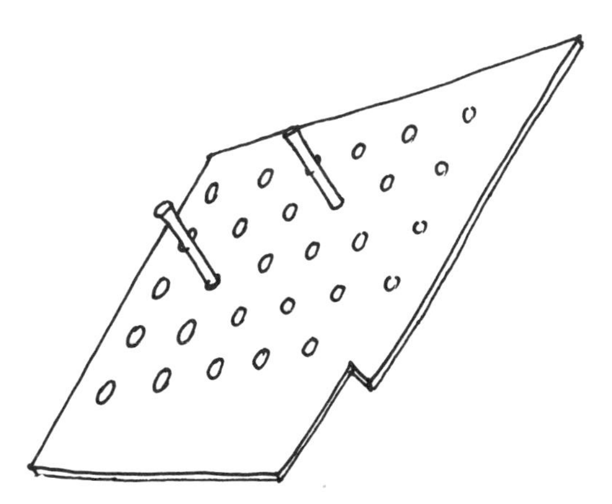

A much neater but more expensive and complicated method of joining components permanently is stripboard. Stripboard is made from a plastic sheet which has a grid of holes in it. Strips of copper are fastened to the back of the plastic so that soldered joints can be made. The strips of copper are called 'tracks' and leads which are to be joined must be pushed through holes on the same track. Sometimes you have to cut through tracks but you will not have to for the moisture sensor circuit described here.

Components can be made to fit quite closely together using stripboard. It is better if they do not touch each other. Resistors can be put in standing on end (see the illustration below) to save space.

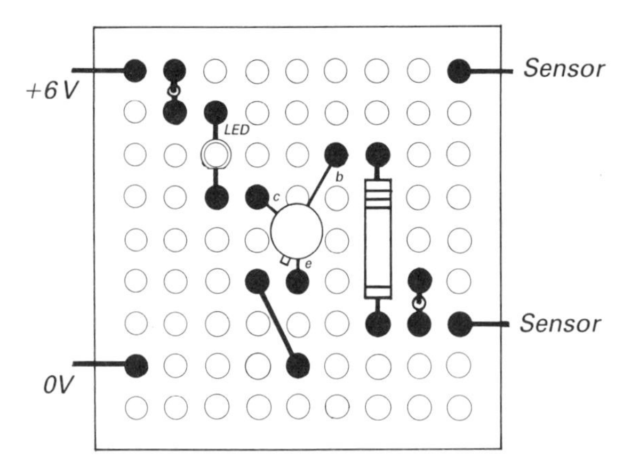

The copper strips are underneath and run in this direction ⟷
The square below shows the actual size of the circuit.
Could you design and make a sensor out of stripboard? You will need to connect alternate tracks (every other one). When a drop of water or a wet finger is placed on the sensor the LED should come on. Read page 80 if you cannot remember what LED means.

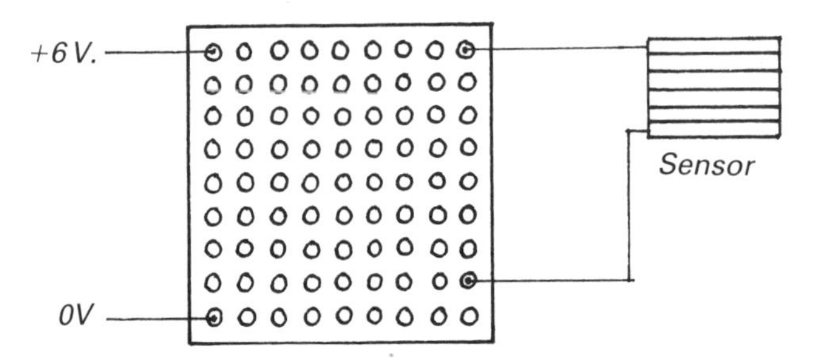

Cardboard pulleys

Decide on the diameter of the pulley and draw a circle on thick card or strawboard. Make sure that the centre is clearly marked. Open the compasses by an extra 3 mm and draw two more circles (mark their centres as well). Cut out the circles so that you have three discs and enlarge the holes at their centres to match the size of your spindles (the rods through the middle).

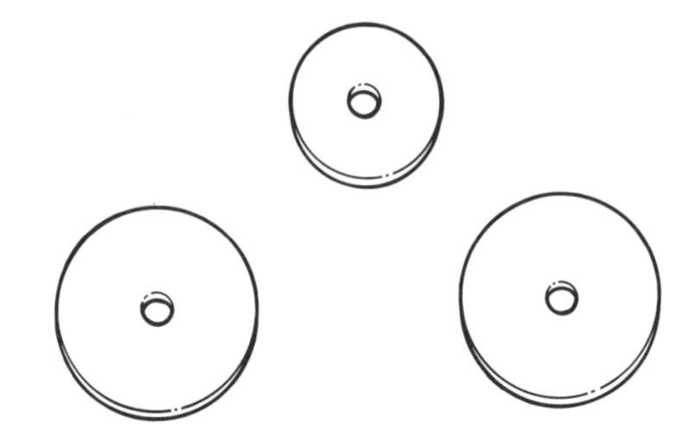

Stick them together using PVA glue and make sure that their centres line up by using a spindle as shown below. Use masses to press the discs together while they dry.

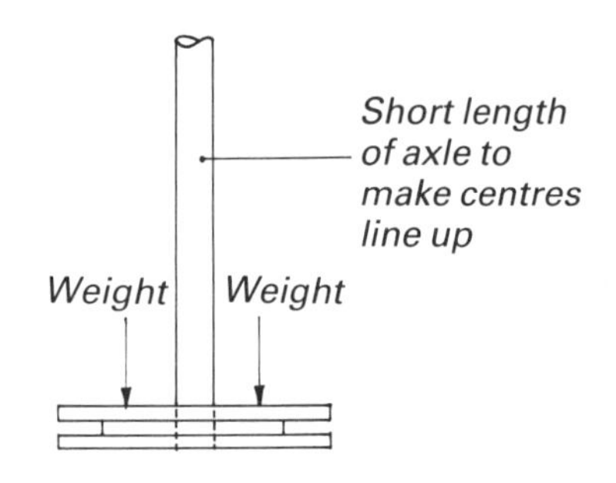

The pulley is attached to the spindle by connectors which can easily be made at school using mild steel bar.

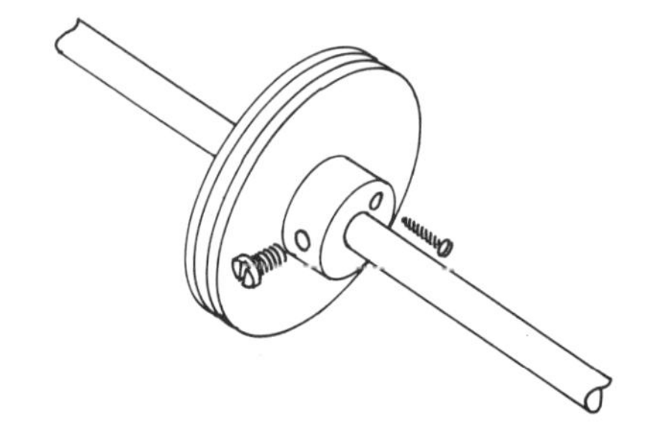

Use rubber bands as belts but find out which sizes are available as you may have to design your pulleys to suit them.

MARKING OUT

The illustration shows tools that can be used for marking out paper, wood, the protective covering on sheets of plastic, and plastics like melamine.
Advantages Pencils and compass leads are cheap, easily sharpened and can be erased (rubbed out).
Disadvantages Leads are easily broken and do not mark metal or other shiny surfaces easily.
Pencil leads are graded in hardness from 6B (very soft) to 9H (very hard). HB is an 'average' grade of hardness. For sketching you may find a B or 2B suitable.

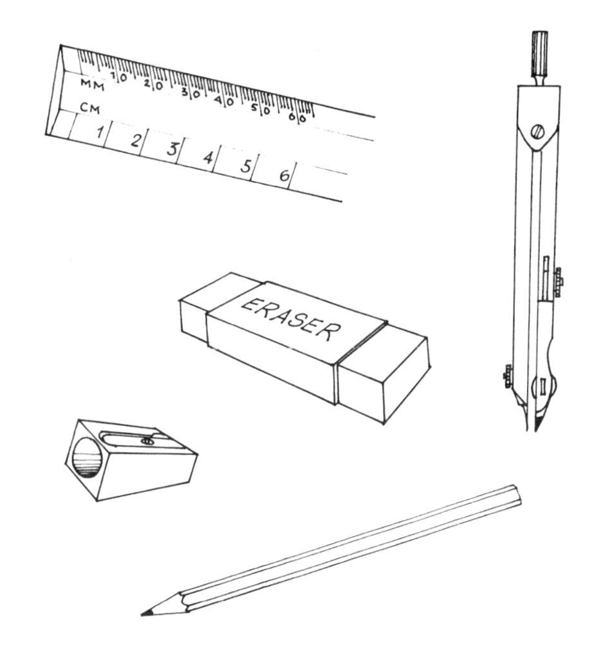

Fibre and felt-tip pens

These pens can be used to draw outlines and emphasise details on drawings.
Advantages They can give a clear line and even colour when working properly.
Disadvantages They may be expensive, difficult to remove, they can sometimes spoil work by soaking through the paper. They will dry up if not looked after.
These pens should not be used for marking out wood. Why?

Look at colour illustrations in magazines. There you will see how colour is used to emphasize certain lines or to give form. 'Quiet' colours are best to use for this purpose.

Scriber or scribe

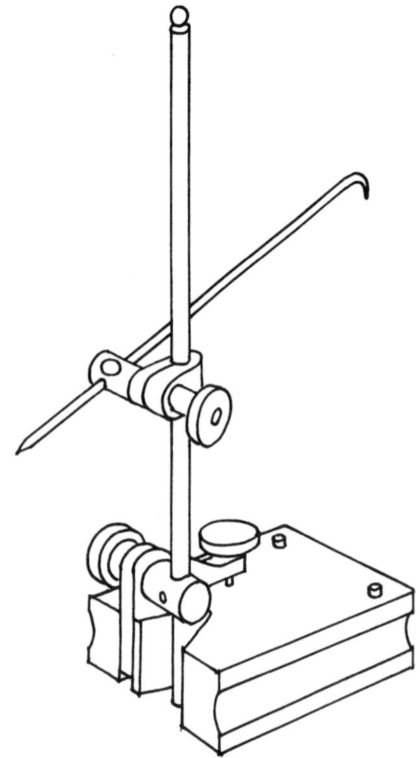

A scriber

Dividers

The illustration shows tools used for marking out metals, plastics and other hard materials. Engineer's blue can be used to coat metals before they are marked out so that lines made with the scriber stand out clearly.
Advantages Scribers and dividers are cheap, easy to sharpen and make thin, accurate lines.
Disadvantages Scribed lines can sometimes be difficult to remove when made on plastics materials.

When marking lines directly onto plastic sheet try using either a B grade pencil, a fine fibre tipped waterproof grade pen or a ball point pen.
Any unwanted lines can be removed later by rubbing with a rag moistened with methylated spirits.

Marking Knife

This is usually used on wood to show where a cut is going to be made.
Advantages This kind of knife is cheap and easy to re-sharpen.
Disadvantages Like most tools with a sharp edge they will also cut the wrong things if not used properly. Mistakes are difficult to remove.

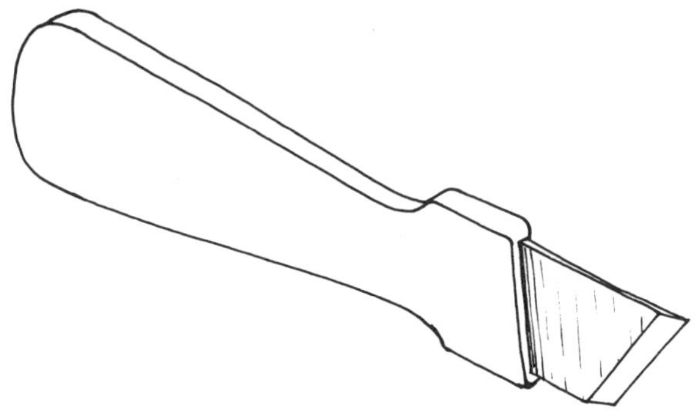

A marking knife; the flat side of the blade presses against a straight edge when used

FORMING MATERIALS

Removing material

Many materials can be cut and formed by using similar methods. A saw cuts when each tooth removes a tiny fragment of material as the blade is pushed forwards.

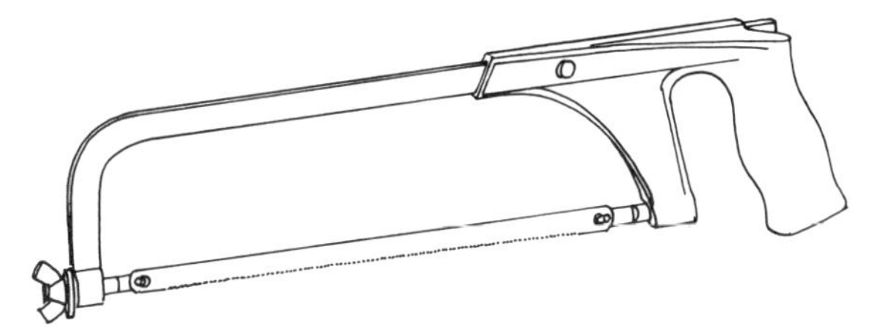

A chisel is a very sharp wedge which can be forced into the material and then remove waste by leverage. Drills and lathes also use the wedge system when cutting.

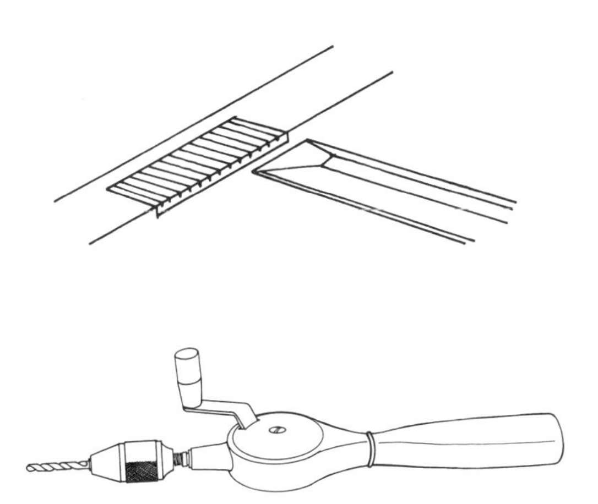

Abrasives have tiny, hard edges which remove pieces of material that are so fine that they often look like dust.

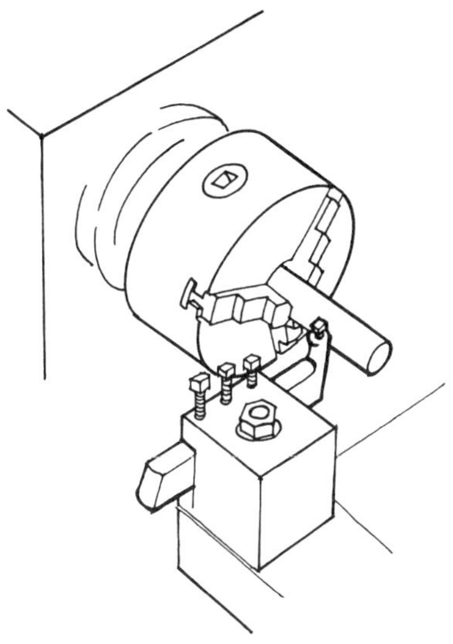

The drawing above shows work held in a lathe chuck. When using a lathe, the work spins round and is cut when a fixed toolbit is pressed against it. When using the drilling machine, the work is still and the bit spins round.

A centre punch mark helps the drill to start in the right place.

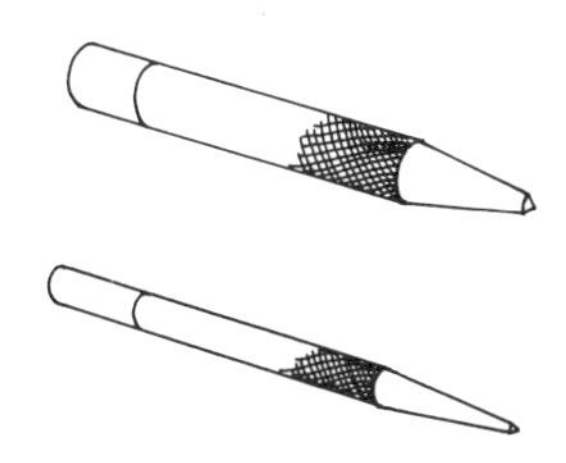

A centre punch (which makes a large dent) and a dot punch below

Scissors and tin snips work by trapping sheet material between their cutting edges and shearing them.

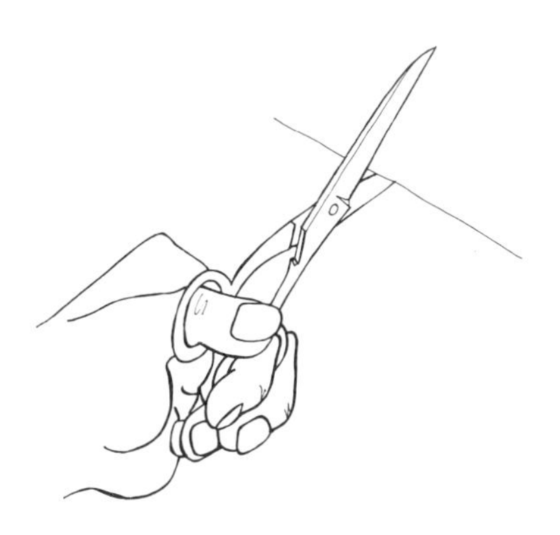

Re-shaping

Materials can be re-shaped as shown in the illustrations. Thick pieces of steel can be formed more easily if they are heated to bright red before being struck. Thin sheet metal can be softened by annealing. Metal which has become hard because it has been hammered too much is annealed by being heated to bright red and then cooled slowly. Plastics materials in sheet form can be shaped on a former as shown on page 39. Straight line bends in thermoplastics sheet can be made using the strip heater shown on page 38. Hot materials can be dangerous, always use the safety equipment provided by the school.

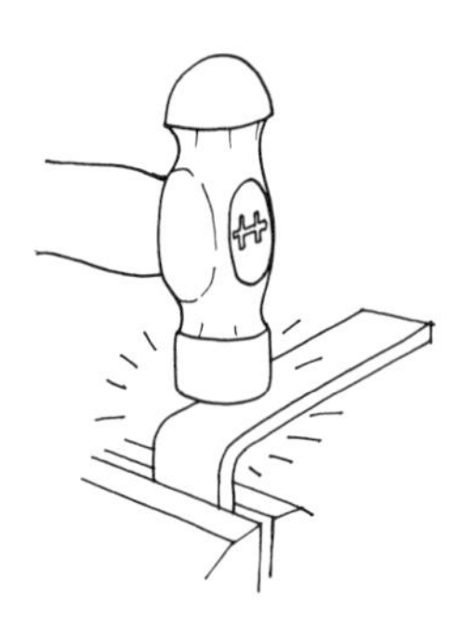

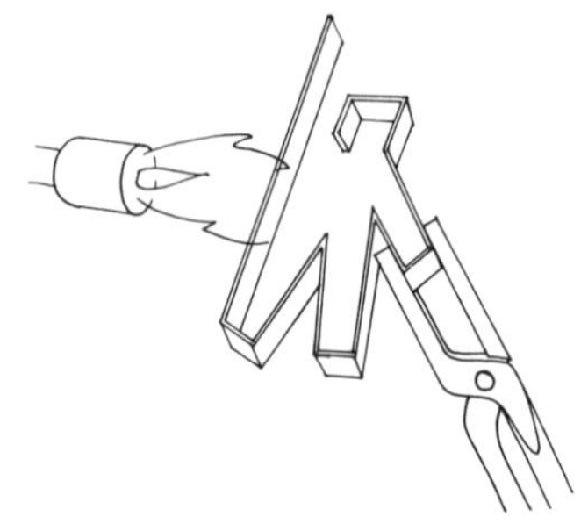

Heating metal using a blowtorch

Laminating

Thin layers of material can be fastened together to make new shapes. Wood veneers (thin sheets) can be glued and then held together in a former while the glue dries. The cardboard pulleys described on page 86 are made in a similar way.

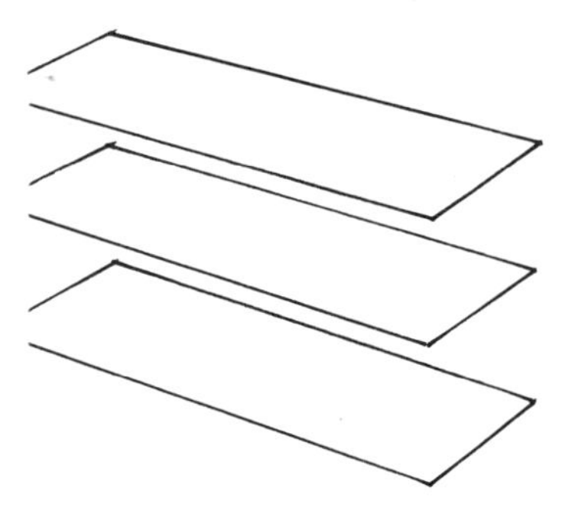

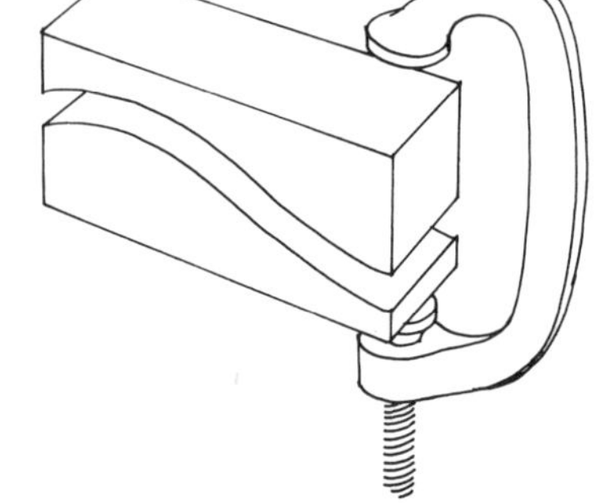

Casting

Metals which have a low melting point (such as aluminium) can be cast in the school workshops (see page 32).
When making a casting in metal, a pattern must first be made which can be removed from the sand easily. The sides of these patterns usually slope (called the 'draw') and have a good surface finish so that the sand will not stick to them. Sand is pressed around the pattern. When the pattern is removed an impression or mould remains. Plastics resins can also be used to make castings. Articles can be surrounded by plastics materials or 'embedded'. Plastic resin sets in a mould as the result of a chemical reaction.

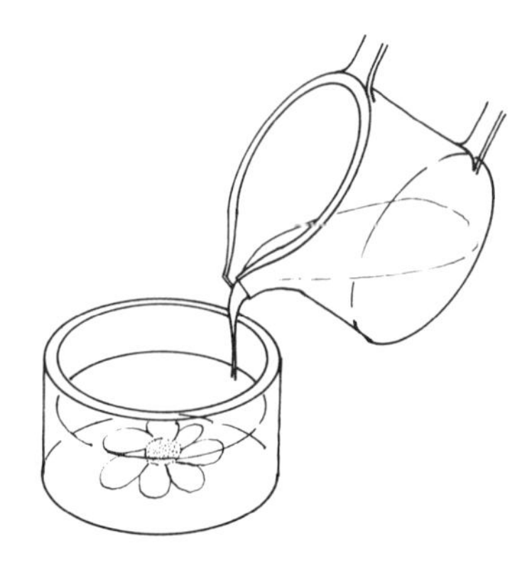

JOINING MATERIALS

Using Adhesives

Water-based adhesives such as wallpaper paste or polyvinylacetate (PVA) glue are very useful as they are fairly cheap and easy to use.

Contact adhesives often have a solvent which evaporates to make the glue stick. Solvents are dangerous—they catch fire easily and can be addictive.

Epoxy resin adhesives usually come in two tubes—a resin and a hardener. The two parts must be mixed together and then spread onto the joint. The adhesive sets when the chemical reaction between the two parts is complete.

Cyanoacrylate adhesives bond (stick) things together by a chemical reaction with tiny amounts of water on the surface of the material. These glues can stick your fingers together! They are also very expensive.

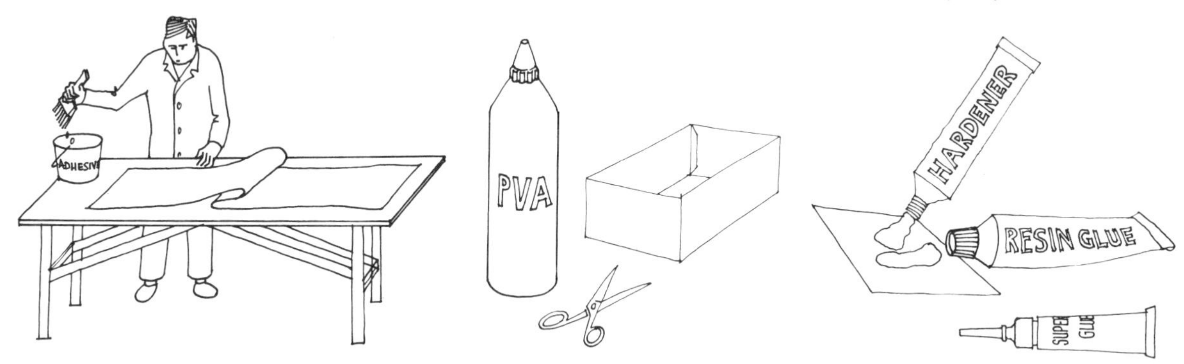

Using nails, screws, nuts and bolts

Nails are used on fibrous materials such as wood. A nail grips because it is driven into the material with such force that friction holds it there. Woodscrews and self-tapping screws have a helix formed into them. See page 61 to remind you what a helix is. The illustration shows how holes should be drilled to help fasten parts together with a wood screw. Nuts and bolts are more effective with washers which make the bolt head or nut grip a larger area.

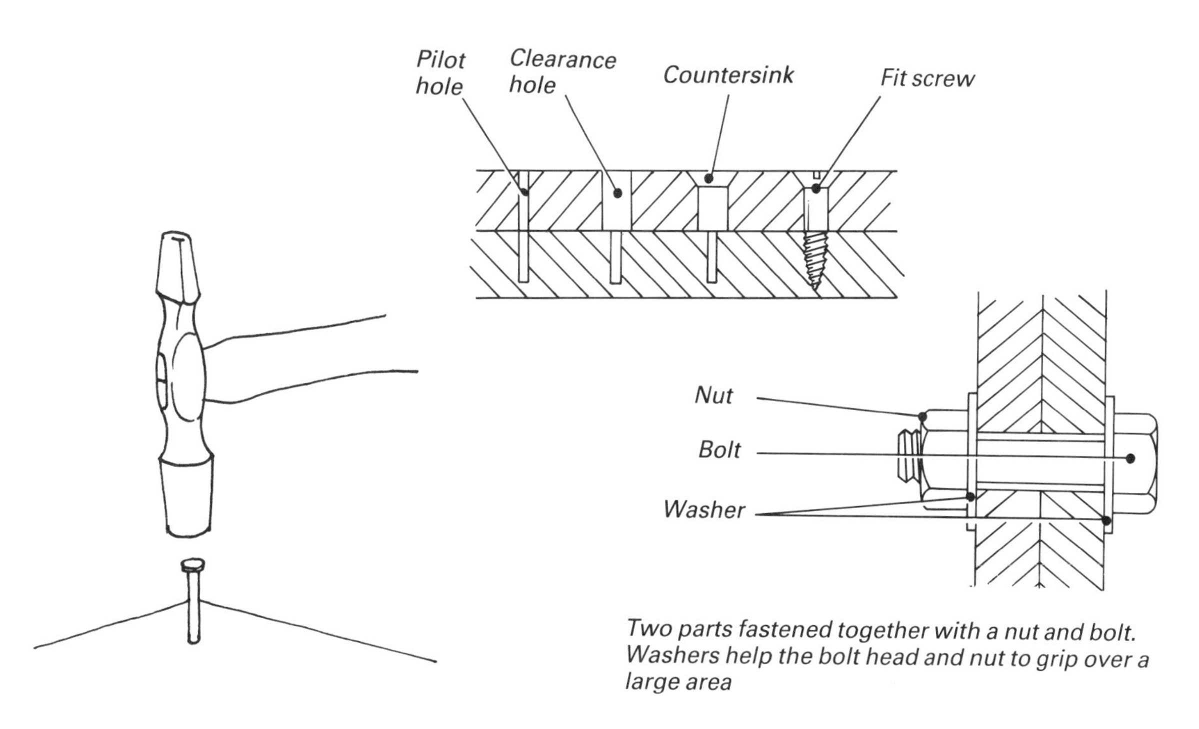

Two parts fastened together with a nut and bolt. Washers help the bolt head and nut to grip over a large area

Wood joints

When glue holds over a large area, joints are very strong. Joints made in wood often have interlocking parts to increase the gluing area.

Joints made in wide pieces of wood are called carcase joints. Framing joints are used in narrow parts.

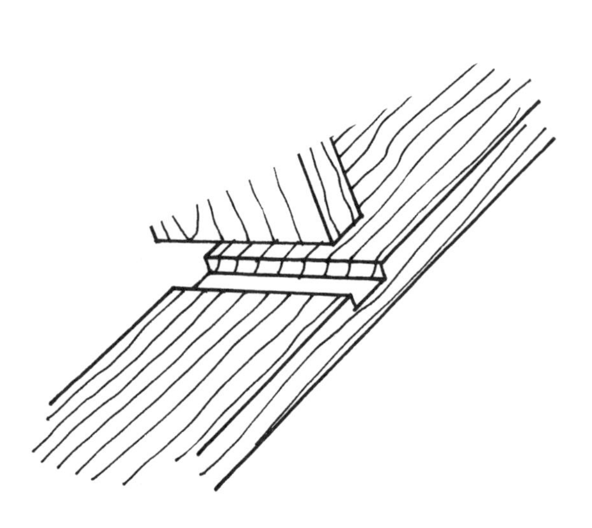

A housed partition joint

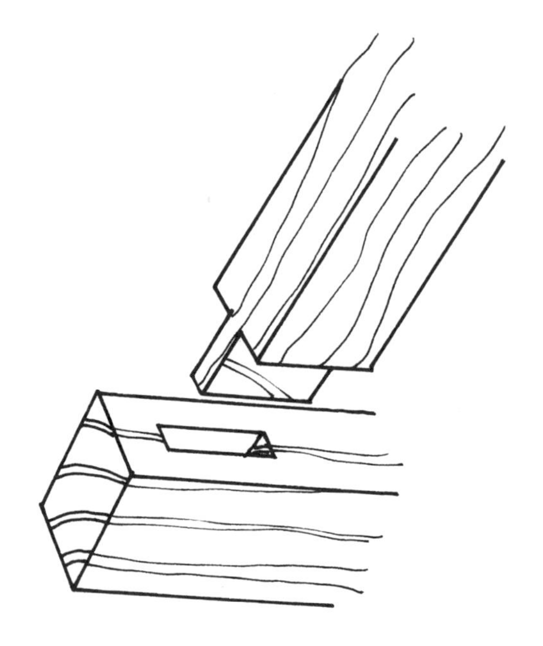

A mortise and tenon framing joint

Soldering and brazing

These methods can only be used for metals and need heat. When metal is heated, oxygen in the air attacks it (oxidation) and forms a coating which will prevent us from making the joint properly. This oxidation can be reduced by a flux which is put onto the joint to help keep the air away from it until the solder or brazing spelter melts. Brazing spelter (a kind of brass) will not melt until the joint is bright red hot. Solder melts at a lower temperature which can come from an electric soldering iron or a blow torch.

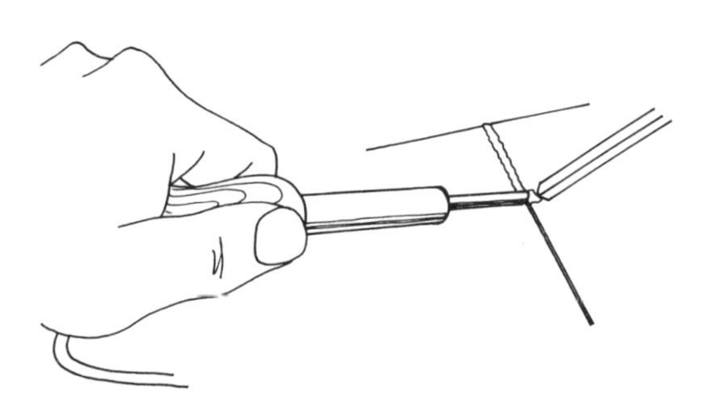

When you have finished using the soldering iron make sure that it is placed in a stand so that it can cool safely

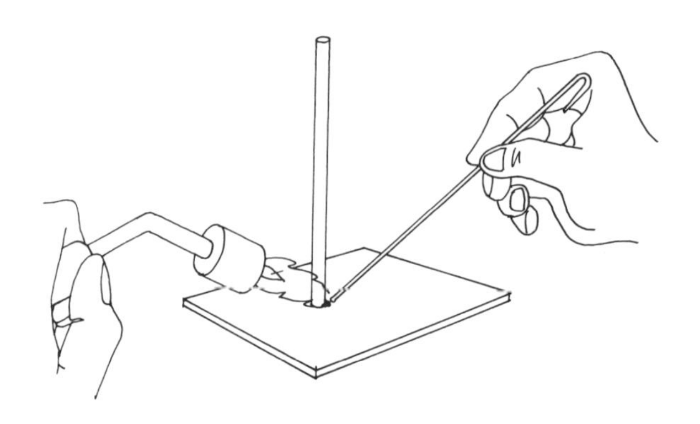

Heat the joint with a blowtorch so that the blazing spelter is melted by the metal and not the flame

SURFACE FINISH OF MATERIALS

Preparing the surface

Some materials, such as acrylic plastic and copper, need very little finishing. Their surfaces and edges will only require polishing with a fine abrasive. Other materials, especially those with absorbent surfaces (like wood) require more work.

The surface of the material is rubbed down with an abrasive to make it smooth and to remove marks. Deep marks and holes which are to be hidden are filled with a paste which adheres (sticks) to the material and can be rubbed down when set.

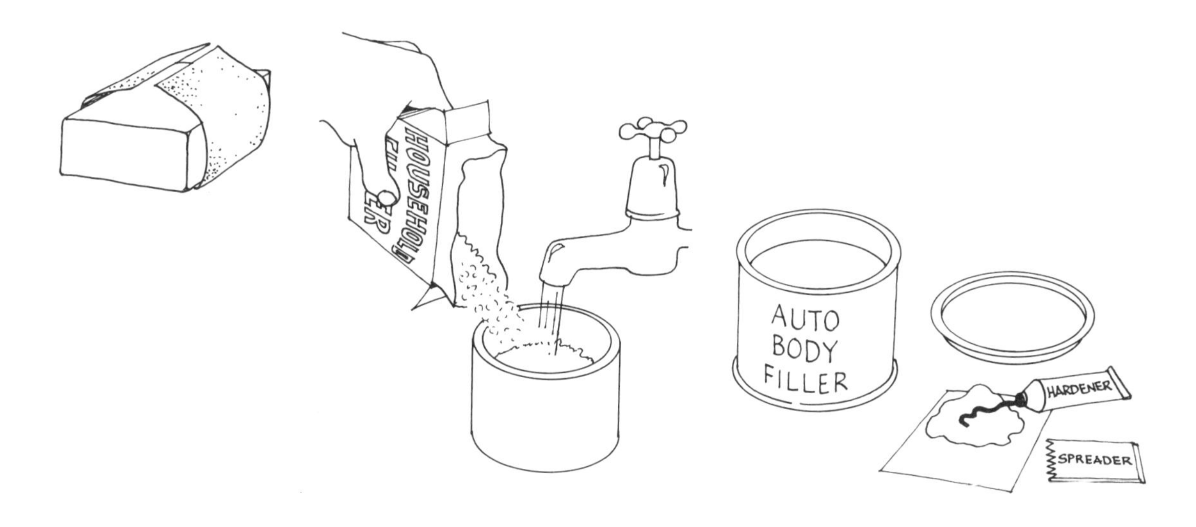

Staining

When wood is to be stained, a mixture of filler and wood dye is used to hide deep marks. This is very difficult to do—it is easier to avoid making the marks in the first place!

Wood that is to go outdoors must be protected by a wood preserver like creosote which often stains the wood.

Painting and varnishing

In schools paint or varnish is usually put onto a surface by brush. A very good finish can be obtained by using spray paint from an aerosol. 'Priming' a surface means putting on a layer of paint which can be rubbed down to cover any tiny marks.

Paint and varnish often raise the grain which makes the surface feel rough when dry. Primer or raised grain is rubbed down or flatted with an abrasive such as fine glasspaper before the next coat is applied.

Several thin coats are better than a few thick ones.

With patience and a lot of care you should get a finish that you will be proud of.

INDEX